TECHNOLOGY CYCLES

CYCLES

and
U.S. Economic Policy

IN THE EARLY 21ST CENTURY

TECHNOLOGY CYCLES
and U.S. Economic Policy
IN THE EARLY 21ST CENTURY

NATHAN EDMONSON

Routledge
Taylor & Francis Group

LONDON AND NEW YORK

First published 2012 by Transaction Publishers

Published 2017 by Routledge
2 Park Square, Milton Park, Abingdon, Oxon OX14 4RN
711 Third Avenue, New York, NY 10017, USA

Routledge is an imprint of the Taylor & Francis Group, an informa business

Copyright © 2012 by Taylor & Francis.

Library of Congress Catalog Number: 2011029347

Library of Congress Cataloging-in-Publication Data

Edmonson, Nathan.
   Technology cycles and U.S. economic policy in the early 21st century / Nathan Edmonson.
      p. cm.
   Includes bibliographical references and index.
   ISBN 978-1-4128-4305-8
   1. Technological innovations—United States. 2. United States—Economic policy—21st century. I. Title.
   HC110.T4E356 2012
   338.0640973—dc23
                           2011029347

ISBN 13: 978-1-4128-4305-8 (hbk)

Contents

Acknowledgments

The author wishes to express special thanks to Malcolm Gillis and Authur Kartman for a number of constructive suggestions regarding the manuscript.

Introduction

The opening decade of the twenty-first century coincided with the immediate aftermath of a major technology-based investment boom and prosperity: the information technology (IT) revolution. Indeed, there was a serious recession in the very opening years of the decade that was attributed to the failure of the so-called dot.com bubble as well as the aftermath of massive overinvestment in an important part of the IT revolution: telecommunications facilities. In the years immediately following this event, growth in the industries most associated with the IT boom subsided, thus giving the telecommunication overinvestment condition of 2000–2001 the additional appearance of having been a foreteller of the general waning of the technology-founded investment boom of the 1990s. Then came the "great recession" that began in 2007, whose proximate trigger was the failure of a bubble in the valuation of housing and related assets. This recession has been distinctive not only by its severity, but also by the weakness of the subsequent recovery. Government responses to the recession have been based on standard countercyclical policy tools that make no allowance for the structural changes in the economy wrought by the IT revolution, and have been largely ineffective. The most pressing question at the outset of the century is "can government macroeconomic measures that ignore fundamental structural changes in the economy maintain economic prosperity at a reasonable level?" The experience to date has not been encouraging.

Many who study economics agree with the proposition that technological change is intimately related to and fosters overall economic growth. The question of how it does this has spawned at least two different schools of thought regarding the nature of the relation between technological change and growth. One of these, which might be termed the "uneven" school, sees the impact of technological change as not uniform through time, as being subject to surges separated by

1

periods of lesser effect. Joseph A. Schumpeter was highly intrigued by a historical tendency for entrepreneurial activity, which drove periods of relatively high growth across wide spectra of industries, to cluster in certain intervals of time, and languish in other periods. While he did not attribute the resulting pulsating nature of entrepreneurial activity exclusively to technological change, he rejected the idea that economic progress could be attributed to simple accumulation of capital stock of the almost same kind.[1]

A second school, the "even" school, emphasizes continual and accumulating technological improvements as being the main movers behind economic progress.[2] The notion of the heroic inventor is largely rejected on the grounds that inventors of record seldom change the direction of technological progress by more than a small amount because their results almost always benefit from the work of antecedents and contemporary enabling technology. They are fortunate in working at a time when the society in which they live is receptive to their inventions and discoveries. This approach has been expanded by a number of studies, which tend to examine technological change in particular industries. The typical conclusion of these is that technical improvement is the result of many small incremental changes, and that there was no conspicuously large change that entered into the total improvement over the historical period of the study.[3]

The even school rejects the idea of dramatically large pulses of technological development influencing growth in short historical intervals. Rather, it encourages the notion that technological progress has occurred uniformly and monotonically through time. Indeed, this seems to be the way much of the economics profession has viewed it, at least since the 1940s. This view has had an important appeal, for if technological change is a slow and steadily evolving background behind financial events, then short-term economic analysis need not bother with it. If this were true, it would be a powerful aid, for it would not be necessary to consider the state of technology flux in macroeconomic forecasting models. It could be safely assumed that a public policy measure would have the same effect whenever implemented, for advancing technology would always be present to absorb liquidity productively, and restoration of strong growth after a financial crisis could be achieved with little more than restoration of financial markets to proper working order plus a possible injection of fiscal spending (pump priming).

The uneven school originated in studies of long statistical series which appeared to exhibit long-term wavelike characteristics. These conclusions are associated with Kondratief, Spiethoff, and others. The even school has tended to dismiss these findings as having more appearance than reality. J. M. Keynes lightheartedly dismissed Spiethoff's system of a long cycle on which was superimposed three shorter cycles by saying that the main utility of this system was to help in reducing long treatises to manageable smaller volumes. Mandelbrot has observed that *ex post* examination of long data samples tends to suggest a Spiethoff-like three-cycle pattern, even when there is apparent total absence of *ex ante* cyclical components in the generating rule underlying the sample. He attributed the appearance of *ex post* cyclicality to long-term serial dependence in the data.[4]

A view of the U.S. economy in the last 150 years provides much circumstantial evidence that changes in the *rate* of technological change do influence economic growth rates. From time to time there are technological developments whose impact goes well beyond any single industry, and an industry-centered approach to the study of technology and its relation to growth is poorly equipped to recognize these effects. When a powerful new technology (or group thereof) affects a variety of industries, the results typically include the emergence of entirely new industries, technical transformations of some existing industries, and the disappearance of other older industries. Any view of technology progress that rests on the assumption of a stable and well-defined structure of industry is unlikely to detect these dynamic aspects of economic growth. This dynamic process has been given a name: creative destruction.[5]

The assumption that technology improvements proceed slowly and more-or-less steadily in time may reflect reality for limited time spans, but it is nevertheless seriously misleading over longer reaches of history. At this point it is useful to adapt a concept from Thomas S. Kuhn: *normal* and *revolutionary* technological improvement.[6] Normal technology advance envisions known technology as improving incrementally and steadily. For example, the Bureau of Economic Analysis official capital stock series basically rests on the idea that worn-out capital is replaced with capital of similar or only incrementally improved productivity characteristics.[7] Normal technological change is predictable, at least conceptually, as it stems from incremental extension of well-established threads of development in the context of a well-defined structure of industry. The tendency for large corporations to concentrate their research and development resources on the improvement of existing product lines is an important aspect of normal technological development.

There is an important difference between revolutionary *science* and revolutionary *technology*: while the former can lead to the latter the time interval between the two can be long. For example, nanotechnology, a *scientific* development of revolutionary proportions, was emerging from the laboratory contemporaneously with the IT revolution of the 1990s. It was not, however, a major contributor to the investment boom that gave economic growth a powerful push in that period of history. Those close to nanotechnology make the strong case that it could underpin a strong growth episode in the twenty-first century.[8]

Revolutionary technological change grows from invention that ultimately brings large-scale obsolescence across wide spectra of industry. It enables *new* combinations of technology that result in dramatic jumps in capital investment for the purpose of applying the new technology, including the replacement of productive capacity that it has rendered obsolete, and provision of infrastructure that the new technology necessitates. Inasmuch as revolutionary technology often comes in the form of novel *combinations* of ideas, it contrasts with normal technology in that it is largely or wholly unpredictable. Technology revolutions impart *exogenous* shocks to economic growth. Far from fitting comfortably into the context of a well-defined structure of industries, they are *disruptors* of the existing industrial order. Technology revolutions spawn entirely *new* industries, increase the efficiency of some older industries, and condemn still other pre-existing industries to economic oblivion.

Technological revolutions are relatively infrequent; there were only three such events in all of the twentieth century. The first spanned the years from about 1910 to 1923 and grew from the electrification of the economy and the growth of the automobile and supporting industries. The second was from 1955 to about 1970, in which a number of new technologies spread throughout the economy. The third was the IT revolution of the 1990s.

While temporally rare, however, technology revolutions decisively influence inventive and innovative activity in longer inter-revolutionary periods of normal technology change. The normal technology that follows a technology revolution in no way can be considered to be a simple continuation of the normal technology that preceded the revolution. Technology revolutions are not overnight affairs; they can last for several decades. When revolutionary technology emerges, it can take time for its possibilities to become widely appreciated. The progress of a technology revolution is regulated by a *learning curve.*

There is no widely accepted index depicting the state of technological development. Therefore, any discussion of such an index has to be in terms of its effects. However, from these effects, it is possible to infer at least some of its characteristics. For example, consider a hypothetical technology index T_t, and its likely statistical characteristics. To suppose that T_t is a steadily (or even monotonically) changing background to conventional macroeconomic analysis tempts one to the assumption that the process generating T_t is essentially Gaussian. This would imply that T_t is independent of T_{t-1}; that is, that a technology revolution is simply a run of results in what amounts to a coin-tossing game. However, this is unlikely, for students of technology history recognize that current and recent values of T_t are related to earlier values of the index; this is the implication of the observation that technology progress rests on past technology developments. Therefore, T_t is a time series having a complex pattern of long-term serial dependence. This suggests strongly that the occasional appearance of an episode of technology change powerful enough to generate a growth-producing investment boom is a systematic part of technical change and not simply an accidental run that can appear in a coin-tossing game, as would be implied by the even school way of thinking. A cyclical pattern in T_t points to the real possibility that differences in the intensity of technological change can affect the relations between standard tools of macroeconomic policy and the rate of economic growth.[9]

Both normal and revolutionary technological change induce investment; but with different implications for growth. Conditions of normal change induce replacement of worn-out productive plant plus net investment in response to perceived need for expanded output. With normal technology, required infrastructure is already in place, and its ongoing maintenance is the only investment that it will induce. In a technology revolution, however, investment in applying the novel technology, in new productive plant to replace productive capacity rendered obsolete by the new technology, in infrastructure needed to utilize the new technology, combine to induce a boom in capital investment large enough to induce a surge in general economic growth.[10]

The 1970s are remembered as the decade of stagflation: a time in which a serious tendency toward inflation coexisted with anemic general economic growth. According to widely accepted economic theory of the time, these two conditions were unlikely to occur simultaneously. Following short but severe recessions in the early 1980s, inflation began to abate. Monetary policy prior to 1980, which had been expansive in the hope of stimulating business investment and economic growth, succeeded mainly in fomenting inflation. During the 1980s, monetary policies reverted to ease following the sharp recession, but with dramatically different effect. Inflation diminished, and economic growth resumed and accelerated after 1990. What made the difference? Naturally, there was interest in this question at the time, but it is questionable that offered explanations were adequate. This is the question that drives this book.

The contemporary explanation was that after 1980, there was a shift in monetary policy emphasis towards controlling inflation. This appeared to be successful in its object, and a period of improving price stability set the stage for an era of generally strong growth that accelerated after 1990 and lasted, with the interruptions of two brief recessions (1991 and 2001) until about 2005. However, it is one thing for monetary policy to create conditions conducive to growth *and an entirely different thing for such policies actually to foment the growth*. Stable monetary conditions supposedly produce growth by encouraging capital investment. Such investment results in expanded productive capacity. To produce growth, however, expanded production based on simply unchanged or slightly improved technology applied in the context of a well-defined industry structure is not sufficient; only investment in technology of sharply superior productivity characteristics will achieve expanded output *per capita*, a necessary condition for growth. If the expansion of capacity is little more than a simple expansion of the preexisting capacity using only the same or slightly improved technology, it is not clear that the process has much growth-inducing potential at all.[11]

As for the disparate implications of easy money in the 1970s and the 1980s, several technologically interesting developments emerged during the 1980s that were not present in the previous decade. One of these was deregulation of ground freight transportation in the United States, which resulted in reductions in the costs of moving goods around. More importantly, however, this was the decade in which the IT revolution was forming after decades of innovation of its basic components, and it was the investment in IT technology that brought the IT revolution to the wide economy in the form of a growth-spawning investment boom. It was the forming IT revolution that gave the economy of the 1980s and 1990s the ability to absorb the rapidly expanding liquidity in the economy productively. Productive absorption of expanding liquidity damped any tendency toward serious inflationary consequence. The monetary policies of the period *enabled* the general upswing, but these were not the *fomenter* of it; it grew from the massive investment in newly available IT technology, especially in the 1990s.

As of 2010, technology change has reverted to the normal. Monetary policies following 2001 were expansive, aimed at encouraging investment, but in what? The aftermath of the IT revolution has included a temporary poverty of productive investment opportunities. The economy was vulnerable to a setback from a shock from the financial sector. This arrived in the form of the failure of the housing bubble, and the result was the great recession following 2007, from which recovery has been anemic to date. The weakened slate of productive investment opportunities has not been adequate to bring a return to vigorous growth. Does that await the next technology revolution?

The technology-founded upswing that animated the economy, especially after 1990, was not the first such episode in U.S. history; it was a repetition of an established pattern that goes back to the earliest days of the industrial revolution. There were three identifiable technology revolutions that affected general growth in the twentieth century. The pulsating character which technology revolutions impart to general economic growth has never received the attention it deserves in the making of government economic policy. The purpose of this book is twofold: (1) to describe in detail how the long technology cycle results in a pulsating character to economic growth; and (2) to show that the weakness of the recovery from the 2007–2009 recession to date reflects a temporary poverty of opportunities for investment in productive assets that follows the growth pulse imparted by the IT revolution.

Notes

1. "The kind of wave-like movement, which we call the business cycle, is incident to industrial change and would be impossible in an economic world displaying nothing except unchanging repetition of the productive and consumptive process." Joseph A. Schumpeter, "The Analysis of Economic Change," *The Review of Economic Statistics* XVII, no. 4 (May 1935). As

reprinted in American Economic Association, *Readings in Business Cycle Theory* (Homewood, IL: Richard D. Irwin, 1951), 7.

2. See A. P. Usher, *A History of Mechanical Inventions* (Cambridge: Harvard University Press, 1954; 1st ed. 1929).

3. For example, see A. Fishlow, "Productivity and Technical Change in the Railroad Sector, 1840-1910," in *Output, Employment, and Productivity in the U.S. after 1800* (New York: National Bureau of Economic Research, 1966). Studies in Income and Wealth No. 30.

4. Benoit B. Mandelbrot, *Fractals and Scaling in Finance: Discontinuities, Concentration, and Risk* (New York: Springer, 1997), 57.

5. This idea had origins well back into the nineteenth century and is associated with the name of Werner Sombart, a German sociologist. It was popularized in the twentieth century by J. A. Schumpeter, and attracted renewed interest during the 1990s. See Hugo Reinhart and Erik S. Reinhart, "Creative Destruction in Economics: Nietzsche, Sombart, Schumpeter," in *The European Heritage in Economics and the Social Sciences*, ed. Backhaus, Jürgen, and Drechsler (Boston, MA: Kluwer).

6. See Thomas S. Kuhn, *The Structure of Scientific Revolutions* (Chicago, IL: University of Chicago Press, 1996). Kuhn introduced the concepts *normal* and *revolutionary* science.

7. The obsolescence associated with normal technological change includes not only technical incremental improvements, but planned obsolescence. Planned obsolescence refers to the deliberate design or manufacturing of a product to fail in some crucial manner after a small number of years so as to stimulate demand for a replacement product.

8. See Malcolm Gillis, *New Perspectives on 21st Century Technology: The Nano-BIO-Info Convergence* (Houston, TX: Rice University Department of Economics). Prepared for presentation at Federal Reserve Bank of Dallas, Houston Branch, April 20, 2010.

9. "It is a peculiar property of most long-memory processes that *seeming* patterns arise and fall, appear and disappear. They could vanish at any instant. They cannot be predicted." Benoit Mandelbrot and Richard L. Hudson, *The (Mis)behavior of Markets: A Fractal View of Risk, Ruin, and Reward* (New York: Basic Books, 2004), 189.

10. This is consistent with what has been known as underconsumption theory. In this hypothesis, a permanent "normal" condition, as used here, leads eventually to a general economic stagnation due to decreasing returns to investment. It also leads to the economy's inability to support full employment. Adherents of this hypothesis included J. M. Keynes, Alvin Hanson, and P. A. Samuelson.

11. This conclusion can be drawn from what has become known as the exogenous technology hypothesis which grew out of a paper by Robert Solow in 1957. See Robert M. Solow, "Technical Change and the Aggregate Production Function," *Review of Economics and Statistics* 39 (1957).

1

A Long Technology Cycle: Outline

*What experience and history teach is this—that people and govern-
ment never have learned anything from history, or acted on principles
deduced from it.*

—G. W. F. Hegel

Origin and Description of the Technology Cycle

Growth in wealth reflects economic growth per capita, and its origin
is the advancement of technology. Technology advancement begins
with invention and progresses with innovation of original invention.
In its first stage, successful innovation transforms an invention from
a laboratory curiosity or a tinkerer's discovery into a useful product,
often one that enables accomplishment of some known task with
improved efficiency.[1] Stage-one innovation is a process that brings
down the cost of applying the new technology to the point where it
is attractive to employ it commercially somewhere. A small subset of
all original inventions that has the potential for accomplishing tasks
well beyond the familiar will be referred to as *prototype inventions*. The
power of the prototype invention derives not only from its enabling
of doing familiar tasks more easily and cheaply, but from the possibilities for doing far more than their inventor/discoverer envisioned by
enabling novel combinations with other prototypes and with already
existing threads of technology. Prototype inventions are the ultimate
source of technology revolutions.

After the first stage of innovation of a prototype invention, there
follows a stage-two of innovation in which the principles underlying
the original invention become applied in novel ways to produce new
kinds of products and productive combinations. Stage-two innovation
brings the results of these combinations into general use by means of
a capital investment surge that underpins an acceleration of general

economic growth. The investment surge begins when the first stage of innovation has brought the cost of applying the new technology down to levels that make its application economically attractive in some major sphere of activity. It expands as entrepreneurs exploit the new technology in combination with other technology threads to produce novel results. The investment that novel combinations induce provides most of the growth force of a technology revolution.

There has been a pronounced tendency for powerful new technologies to appear in bunches in comparatively short spans of historical time.[2] When this happens, the stage-two innovation processes create an unusually rich slate of investment opportunities. A powerful stage-two innovation episode is typically marked by higher-than-usual productivity gains across a wide spectrum of industry, the appearance of entirely new industries, and disappearance of older industries that were based on technologies obsolesced by the new technologies. The investments that achieve these changes in the economy at large engender a period of unusual general prosperity.

Episodes of strong technology-founded growth are separated by sometimes lengthy intervals of normal technology change in which technology progresses more slowly, or not at all. Inventive activity focuses on improving the combinations that resulted from the most recent technology revolution. It is in these intervals that the economy becomes vulnerable to serious problems, such as low growth, serious recessions, inflation, and combinations thereof, which are often triggered by financial market disturbances of various kinds. This alternation gives long-term economic growth a pulsating character. The twentieth century was marked by at least two powerful technology-based prosperities, one lasting from about 1910 to the mid-1920s, and the other from 1980 to about 2005. The general prosperity of the first two decades following World War II also rested on a strong technology foundation. In these examples, the prosperous upswing eventually collapsed. In the early-century experience, the collapse was into the great depression; the postwar prosperity degenerated into the stagflation of the 1970s; and in the most recent case, the collapse was into the recession that began in 2007 and which now presents uncomfortable reminiscences of the 1930s.

The pattern of economic expansion from the onset of a technology-founded upswing to its peak is not typically a uniform expansion. Upward progress can be interrupted by shorter-term financial crises of varying magnitude. The technology-founded cycle can be viewed as a basement over which the world's financial markets operate. The most important thing to remember in the relationship between the technology cycle and recurring superimposed financial events is that if a financial crisis occurs during the upswing phase of the technology cycle, the recession will generally be far less severe and recovery more robust than if it coincides with the end of a technology cycle. If it occurs after the end of a technologically founded upswing, then the possibilities range from a period of stagflation at best to something akin to depression.

The reason is that if the fund of technology-based investment opportunities has not been fully exploited, then once the financial problems are resolved, the technology-based expansion can resume. If, however, these investment opportunities have become fully exploited, the recession is likely to be more severe and prolonged than would be the case if there remained a fund of investment opportunities to exploit.[3] At the heart of the problem is that recovery of vigorous growth requires a new round of technology-based investment opportunities. Absent this revival, investment is not a fomenter of growth, for it is only the means by which the benefits of new technology are spread to the economy at large.[4] Indeed, there is a tendency historically for declines into serious recessions to be triggered by financial crises at or after the end of a technology cycle, and for contemporary explanations of the downturn to focus on the financial crisis and ignore the reasons underlying investment failure. The year 1929 is an outstanding example. The anemic character of the recovery from the recession of 2007 suggests that the post-1929 pattern is repeating itself.

The Cycle: Upswing and Failure

Onset and Upswing

The investment upswing derives from the stage-two innovation of a group of powerful prototype inventions. An increase in private investment has two effects: the general economy expands through the operation of the investment multiplier; and the investment in capital with productivity characteristics superior to that which it replaces brings a parallel expansion in the economy's capacity to produce goods and services, Γ (Figure 1.1). The relationship between the investment surge and capacity expansion is of interest as some past writers on economic cycles regarded capacity as a constraint that limits economy's expansion potential. Generally speaking, such models depict cycles of much shorter duration than what is envisioned here. In these, the cycle duration is too short for the recycling of resources that became unemployed during the recession.[5] In the long technology cycle, the investment surge brings new capacity to produce goods and services even as it renders older capacity obsolete. The duration of the upswing is sufficient for the recycling of resources rendered obsolescent by the new technology.

As illustrated in Figure 1.1, X (Gross Domestic Product) increases *relative* to Γ, thus diminishing the vertical gap between Γ and X, or

Figure 1.1
The Technology Cycle

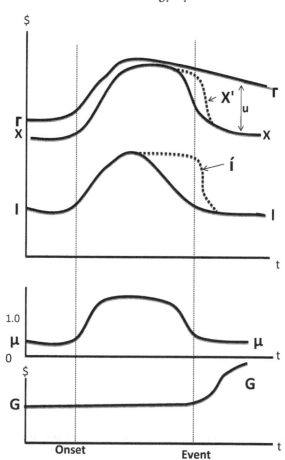

the unemployed resources of labor and capital available to support expansion. This gap is denoted by the symbol u in the figure. Labor and capital resources for the support of the upswing in addition to whatever unemployed resources were available at its outset come from the recycling of productive facilities obsolesced by more productive new capital. A powerful long-term investment surge thus provides its own resource support as it progresses. A weak investment surge will not always create sufficient capacity to assure avoidance of inflationary pressures and other problems that arise from increasing scarcity of resources to support expansion, but historically, the pace of economic expansion has often proved to have been sufficient to allow re-absorption of obsolesced resources without serious immediate problems of a capacity constraint.

The upswing phase of the technology cycle is driven by conditions in which the *expected* return to capital investment *exceeds* the cost of capital.[6] Of course, the more that is invested in the exploitation of opportunities stemming from a particular thread of technology, the more the returns of such investment will tend toward the cost of capital. However, a group of powerful prototype inventions is likely to spawn *multiple* strains of innovation, and the investment opportunities implied by combinations of these mean that for a number of years at least, expected investment yields will exceed the cost of capital.

The idea of the preceding paragraph clearly borrows from the notion of internal rate of return (Irr). Irr is a tool commonly used in capital budgeting in the individual firm. The capital budgeting problem arises when a company faces more potentially profitable investment opportunities than it can finance consistent with the health of its balance sheet, and wishes to identify the most profitable subset of these opportunities that fit within its capital budget and tolerance for risk. Irr can be derived by solving the equation

$$I_k = \sum_{i=1}^{n} (NCF_{i,k} / (1 - Irr_k)^i) \qquad (1.1)$$

for Irr_k, where I_k represents the k-th investment opportunity, NCF_i is expected net cash flow from the k-th opportunity, and n is the expected life of the project. The idea is to solve for Irr for each of a number of the available opportunities so that the most attractive subset of these can be identified for funding.[7]

The extension of the Irr concept to the full economy requires some conceptual modifications. The most important of these is that when new technology has presented an array of investment opportunities, the problem ceases to be a capital budgeting problem from the standpoint of the full economy, for there is no reason for supposing that the entire economy is limited by a capital budget analogous to that of a single firm. The progress of the upswing is limited solely by the learning curve. Past writers on business cycles have cited over-all productive capacity as an upside constraint on the expansion. A capacity constraint makes sense only in the short term. An upswing that is supported by expanding possibilities growing from a slate of new prototype technologies can last sufficiently long that resources in industries obsolesced by the new technology can be recycled to support the growth.

The vertical axis in Figure 1.2 has rate dimension. The CC curve represents cost of capital, which is heavily influenced by market interest rates. The Irr curve is expected Irr to capital as applied to the full economy. The term Y_e is the excess of yield to investment over cost of capital. The surge in Irr above the cost of capital reflects the economy's exploitation of the investment opportunities created by the appearance of new technology. If the new technology is the result of a particularly powerful cluster of new prototype technologies, the vertical gap between the Irr curve and the CC curve will be greater and longer lasting than if the new technology grows from only a single thread of invention and innovation. The eventual decline of Irr to

Figure 1.2
The Technology Cycle: Upswing

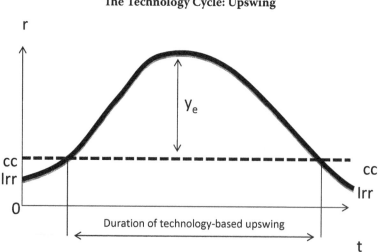

CC and below reflects that, eventually, the markets for the products of the new technology will become saturated. At such time, the new technology is novel no longer, and investment in it has progressed to the point at which new investment has ceased to replace dramatically less productive technology, but only replaces technology of similar productivity character.

Even if established companies shy away from the high degree of risk that can be involved in the development of novel technology, there is nothing to prevent new firms from taking up the ideas, and in some cases, prospering with them. Indeed, upswings growing out of major new technology tend to involve entirely new industries populated by new companies. This is a usual pattern in technology-based upswings.

The expansion phase of the technology cycle receives strength from what economists have termed the multiplier effect. The multiplier concept summarizes the process by which a dollar of capital investment gives rise to a number of follow-on expenditures, each of which constitutes an element of income for someone. The multiplier effect is strong during a technology cycle upswing because investment in new technology characterized by improved productivity creates productive wealth. Obvious examples include automotive-related industries in 1900–1923 and IT-related industries in 1980–2005. Emerging and rapidly growing new industries absorb such unemployed resources as existed prior to the upswing plus resources rendered obsolete by the new technologies. The increased productivity of the new capital reflects that resources are used in a highly efficient manner. Figure 1.1 illustrates this in the middle panel, wherein the multiplier, μ, rises during the upswing of investment, I.

A technology-based upswing, as already noted, can occur contemporaneously with a number of other possible events which result in alternating recessions and prosperity of shorter duration than that of the technology upswing and are superimposed on it. There has been a tendency in the practice of short-run economic forecasting to regard all technology-related developments as exogenous events to which the economy reacts quickly, or perhaps does not at all react during the forecast period. This is unfortunate for two reasons. First, the condition of the technology cycle has much influence on the severity of recessions. It profoundly affects the manner in which macroeconomic policy measures, which a short-run forecast probably did consider, affect the economy. To neglect it is to rob economic theory of much predictive potential. Second, there may be nothing quick in the economic adjustment to new technology because it takes time for economic actors to recognize and act on opportunities provided by it. Stated differently, conventional short-run forecasting makes little or no provision for a "learning curve" effect on economic decisionmakers.

As for the exogeneity of technology events, most such are generated within the economic system and should be treated as endogenous. The only truly exogenous technology events are the invention/discovery of the prototype inventions themselves.[8] A prototype invention in being, however, exerts a powerful influence over the direction of ongoing inventive and innovative activity because such activity works to develop the wider possibilities of the prototype. There is no reason why ongoing development of a known prototype principle should be an exogenous surprise. This is the new normal technological progress.

The Peak

The peak of the technology cycle occurs when the slate of investment opportunities that grew from stage-two innovation of a group of prototype inventions begins to diminish. At the peak, the economy will have achieved a low rate of resource unemployment. The theory of acceleration as typically presented to lower-division economic students holds that as soon as the rate of expansion of X diminishes, investment will begin to decline, as illustrated by the solid line in Figure 1.1. Classroom theoretical exercises assume that the economic actors who make the investment decisions are acutely aware of and instantly responsive to all market conditions that might affect their decisions. However, such assumptions rarely if ever reflect actual conditions. First and foremost among the reasons for this is that to decisionmakers, there may not be immediate obvious reason to curtail investing. Consider what has been taking place up to this point in time. Capital investing in technologies regarded as novel has been rewarded by higher-than-normal rates of return. Such returns reflect the expanding opportunities afforded as applications of technology that stem from a group of especially fecund prototype inventions. It is the nature of these opportunities that they spread through the economy in ways not foreseen by the original inventors of the underlying technology. Of course, hindsight informs us that the bases for hoping for extraordinary rates of return diminish as the various technologies become commonplace, but who among those who have lived through a technology upswing has the crystal ball that says the good times are fading at the precise instant the fading process actually begins?

Low-resource unemployment rates are as likely as not to be interpreted as a sign of continuing prosperity rather as a warning of impending collapse. Suppose investment continues at a high level, as illustrated by the Í (dotted) curve in Figure 1.1. This creates an illusion of general economic activity, X, at high levels. The drivers of the dotted I and X curves are largely psychological. While high levels of expenditures on assets may be able to support continued high general prosperity for a time, their power to foment growth has diminished. What is changed is that the quality of the acquired assets has deteriorated. Before the peak, assets acquired tended to be productive, but postpeak acquired assets are increasingly preexisting (therefore tradable) assets of various kinds: e.g., commodities, securities, and real estate. As opportunities for productive investments become exploited, yield-seeking increasingly takes the form of speculation.

Figure 1.3 illustrates the process by which liquidity pursues yield even after the pool of the most productive investment opportunities approaches full exploitation. The reader will recognize Figure1.3 as a variant of Figure 1.2 to which has been added the curve Y_c, which represents yield accruing to speculation in some commodity or other asset for which valuations are perceived to be rising.[9] In this figure, liquidity pursues productive assets until the return on them falls below that on alternatives. When the Irr curve falls below the Y_c curve, liquidity will follow the Y_c curve. Inasmuch as rising Y_c depends on rising valuations of the underlying asset and often little more than that, dogged pursuit of yield results in deteriorated quality of assets acquired. At this point, "investment" no longer yields the means of increasing goods and services output *per capita*.[10]

The Downturn

Why must the technology boom collapse? One can reason that the conditions relating to full employment should be maintainable with a combination of sensible private investment activity and inspired public policy. To suppose this, however, confuses "should be" with "is," which can be very different. Consider what can happen from the peak of X (see Figure 1.1) while low-resource unemployment is being maintained with the aid of private investment well beyond the point at which a simple accelerator argument would hold that investment should have begun to decline, especially when monetary authorities maintain generous amounts of liquidity in the economy during this phase of

17

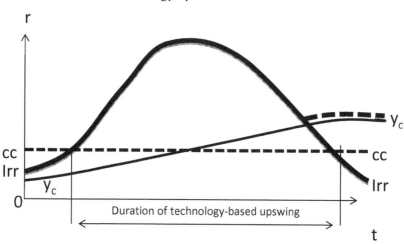

Figure 1.3
Technology Cycle: Pursuit of Yield

the cycle. The possibilities include general inflationary pressures and asset price bubbles, as investable funds seek high returns in the face of vanishing technology-based productive opportunities.

In general, the period of very low-resource unemployment rates following the peak in a technology-based upswing is one of high vulnerability for the economy, and contemporaneous lack of appreciation for this vulnerability increases the surprise at the apparent suddenness of the collapse when it occurs. One can argue the point of which *proximate* events trigger inflation or the bursting of an asset bubble, but the basic problem is fueled by easy credit and a *poverty of opportunities for productive investment*. A collapse in the valuations of commodities and goods heretofore expanding, possibly in bubble conditions, takes place.

Prof. Hyman P. Minsky proposed a financial market cycle mechanism that is of interest when studied in juxtaposition with the long technology cycle. In this system, a build-up of debt relative to incomes, due to various causes inherent in capitalist financial markets, renders financial markets increasingly vulnerable to setbacks. Minsky divided borrowers into three groups. In the first, termed "hedge borrowers," borrowers are able to pay both interest and principal paydown on their loans. The second group pays only interest on its mortgage debt (but does so reliably) and therefore requires additional financing in order to pay back principal. Such borrowers rely on appreciating value of their homes to supply the financing for ultimate payback. The third category (which Minsky termed "Ponzi borrowers") are borrowers who lack the means to service interest—let alone principal—on an ongoing basis, and can borrow only with the lender's conviction of continued appreciation of the value of the asset used as collateral, such as housing. As a period of flush liquidity develops, aggressive pursuit of yield encourages a diminution of lending standards, with the result that the proportion of borrowers in the Ponzi group increases, sometimes enough to render the financial system vulnerable to a reversal in price appreciation of assets used as collateral, whether it be housing (2007) or stock valuations (1929). When this occurs, the overall financial market collapses, for the reaction affects *all* borrowers. When credit availability begins to recover, due possibly to actions by the monetary authority, recovery of the wider economy may lag, for lending standards applied to all borrowers have become stringent.[11]

Asset bubbles happen all the time, but the intervals following the culmination of a technology-based investment boom are times especially prone to their development. When an asset bubble develops while the investment boom still has life, recovery is relatively rapid, and the economy resumes the investment boom. When the technology basis for quick recovery is absent, however, recovery can be weak and prolonged. Moreover, conditions immediately following the end of the technology cycle, especially a high degree of liquidity in the economy, are tailor-made for a progression toward a high proportion of Ponzi borrowers among total borrowers.

Tightening credit conditions are often *perceived* as the cause of the collapse of asset bubbles. They are definitely an aggravator of it, as lending institutions attempt to forestall insolvency by tightening credit availability *after the collapse*. When personal consumption has become supported by consumer credit,[12] as in the years prior to the 2007 collapse, debtors are forced to reduce their consumption when the burden of debt service rises relative to current income. A decline in the value of housing assets relative to mortgage debt seriously impacts consumption, for consumers who had previously counted on rising home values to keep their finances sound find their ability to consume reduced.

Private investment is a casualty. Business organizations that had been willing to venture money into projects which they hoped would yield reasonable or better-than-reasonable returns review and cancel projects whose riskiness appears enhanced in circumstances of reduced revenue, cash flow, and access to outside financing.[13] All of these developments spread the downturn to the economy at large. On Figure 1.1, this is marked by the vertical dotted line labeled "event." The event is typically a financial crisis that, among other things, forces realism on all actors in the economy by highlighting the financial excesses that have developed during the preceding period of apparent prosperity.

One problem in the collapse is that the financial crisis will attract most of the attention for explaining the collapse. The crisis completely diverts attention from the role played by the end of the technology-based boom. This is a problem because the end of the technology cycle implies a temporary poverty of profitable investment opportunities that attract private investment. The absence of such opportunities means that an early revival of strong economic growth is unlikely, a condition that ought to be considered by government in its reaction to the financial crisis.[14] Once the crisis is under way, the most the government can hope to accomplish is the restoration of a normal operation of the financial markets; but restoration of vigorous growth absent a technology basis for growth is beyond the government's capability.

Capacity

As shown in Figure 1.1, capacity, Γ, does not decline with the collapse in Y; it cannot, at least in the short run. Some of the physical plant capacity that was put in place during the technology-based upswing becomes redundant after the downturn, but continues to exist and to affect the earnings of owning firms. These can in time reduce their operating costs by concentrating production in fewer facilities. This explains what may seem a curiosity: the rise in productivity measures that took place in the depression of the 1930s. Under depression conditions, producers resort to layoffs of employees, but at the outset of the downturn, they are not able to reduce costs in proportion with the decline in revenues because some costs are fixed in the short run, such as depreciation on plant facilities made redundant by the downturn. One has to remember the economists' definition of the long run as that period in which all costs become variable costs. As depression-like conditions persisted for years on end, producers were increasingly able to reduce costs even more than in proportion to the reduction in their revenues with the result that the denominator of the productivity ratio eventually fell more than in proportion with the decline in revenues.

What happens to redundant productive capacity in a prolonged recession? In such conditions, revenues and profits tend to be down. Financial write-offs of productive facilities make profits even worse than they would be because of reduced revenues. Hence, there is some tendency to keep shuttered facilities on the balance sheet. In many cases, facilities can be kept physically in existence at low cost—with a skeleton caretaker crew—and could be restarted when business revives. Such a strategy makes sense when it is considered that managements generally have no reliable way to perceive how much longer the difficulties will last. Much capacity that had been shuttered in the depression of the 1930s was available to support war production in the 1940s. Decisions to mothball rather than scrap unneeded capacity tended to favor keeping those facilities embodying the most recent technology in production.

Income Distribution: A Downside

While the effects of a technology revolution are clearly positive for the economy as a whole, they are not uniformly distributed among the population. People least affected by the revolution find their incomes growing more slowly than the incomes of those who are in some sense direct beneficiaries of the revolution. The social problems that can grow from a worsening income distribution may not be apparent as long as the upturn proceeds, but can become so following a collapse into recession. A worsened income distribution problem can manifest itself in the form of voter appeal of income redistribution schemes promised by politicians. These typically involve the use of income tax policy proposals to tax the "rich" disproportionately. Such schemes, when implemented, are highly complex in their effects and can themselves be impediments to cyclic revival.

Persistent Unemployment: A Downside

Persistently high unemployment is a greater immediate downside of a technology revolutionary episode. The reason is that a technology revolution profoundly alters an economy's productive paradigm and therefore can bring serious changes in labor markets. The spread of revolutionary technology, among other things, teaches employers how to operate with fewer hands. This enables employers to reduce costs in the face of declining revenues to a greater extent than would be possible had there been no essential change in production techniques. Rising unemployment in a postboom recession thus contains a large structural element. Because of this structural character, unemployment does not respond to countercyclical government policies, such as fiscal stimulus and monetary ease, that presuppose that high unemployment is mostly cyclical in nature.

Revival

The only way by which vigorous economic growth can revive fully is with the support of a slate of investment opportunities created by a powerful stage-two innovation process of new technology: a new technology revolution. The role of improved technology is crucial, for it ultimately is the sole basis for a significant expansion of the output of goods and services *per capita*, the very heart of the concept of growth. Recovery from a recession that follows the end of a technology-founded boom can be weak and prolonged if there is no immediate technology basis for a growth revival.

Absent this condition, there are several possible ways in which a *limited* recovery may be fomented by government policy measures, which will succeed or fail to the extent that the measures do not prolong the downturn, such as by creating an atmosphere of uncertainty for private business or by crowding out private access to financing. The immediate problem as the economy sinks into a serious and general recession is to restore its normal financial operations to a level approximating that which existed just prior to the collapse. This necessarily means that markets have to resume their normal functions without the bubble conditions that may have existed prior to the collapse and whose failure probably contributed to bringing about the collapse.[15] There are several possible sources of action to encourage this kind of recovery: the private sector, the government, and the monetary authority.

The private sector does not appear to be a likely source of immediate succor despite its role as the powerful engine of growth once the technology upswing emerges. It faces several problems. The greatest of these is that the field of lucrative investment opportunities on which the preceding technology-based boom fed has become temporarily exploited. In exploiting these opportunities, business concentrated most of its resources on producing, improving, and marketing the products that grew from the stage-two innovation of a group of fecund prototype inventions, and away from original invention.

Its new product pipeline is empty or underdeveloped. While there may exist other prototype inventions in stage-one innovation, developing these can be costly and risky. The necessary stage-one innovation does not progress well in the face of reduced profitability and cash flow, and limitations on access to financing, conditions brought on by the general recession.[16] The second is that, as in the present situation, the general collapse of financial markets impairs access to financing even for necessary day-to-day working capital needs of many companies.

This is not to say that the private sector has no role in initiating a general revival. It is the private sector's exploitation of new technology as it emerges from stage-one innovation that begins what can be the beginning of a technology revolution. Whether or not a full-scale technology revolution grows from private investors' early investments in a new technology, depends on the potential of the new technology. The crucial point is that the technology of high potential has to be present, at least in nascent form, for the private sector to undertake its stage-two innovation. A technology revolution is an essentially private sector creation. The prototype inventions that make it possible are mostly unpredictable.

If established companies' new products pipelines are temporarily empty, can new companies take up the slack? The answer is probably not in the earliest days of the crisis. New companies are likely to have been at the heart of the preceding technology-based upswing; indeed, new technology often means new industries in which startup firms are well represented. However, at the end of the technology-based upswing, the growth basis for the new industries has diminished. The slate of technology-based investment opportunities has become fully exploited for *all* companies, old and new. Moreover, the adverse impact of nonfunctional financial markets may fall on small companies proportionately much more than on large, established companies.

Governments can take a number of actions at the outset of a very serious recession, but these vary as to effectiveness. These include stage-one development of technologies originating in defense and other government-supported basic research, the maintenance of tax and legal conditions that encourage entrepreneurship, and stimulus spending. The first two of these have proved highly effective in promoting the revival of technology-based growth, but both are long term in nature. Neither offers a "quick fix" for the immediate recessionary conditions. The third, stimulus spending, promises a quick fix, but is largely ineffective. Moreover, it is widely suspected of inhibiting such private sector revivals as may develop by "crowding out" private access to financing sources.

The problem with stimulus spending is that its effectiveness rests on the assumption that private investment and government spending are somehow interchangeable as fomenters of economic growth. However, this assumption fails to recognize what is happening in the real economy during a financial crisis: resources are being redirected from mature activities and into emerging sectors, but these may not be identifiable at the outset of the recession. Private investment promotes growth only when it implements new technology that enables increased output of goods and services per capita. Absent such new technology, the investment that takes place does not have much growth-inducing potential, for not only is its quantity sufficient only to satisfy the limited needs of normal technology investment, but its quality is little more than what emerged from the most recent technology revolution. Government "investment" cannot deliver growth because it typically is pursued as a countercyclical policy only under conditions in which there is a paucity of productive investment opportunities for either private or government investors. Government expenditures on infrastructure are unlikely to bring about more than minor improvements in a country as well-developed as the United States.

This leaves the Federal Reserve System. Monetary policy unaided by an underlying technology push, it should be emphasized, *lacks the power to reignite vigorous growth.* Easy money policies following the peak of a technology-based upswing do not foment reignited growth, but do create a breeding ground for the development of bubble-like conditions and for the prospering of financial frauds, and the combination of failing bubbles and collapses of fraudulent schemes accompany and aggravate a financial crisis. The most constructive role the Fed can play is to recognize the top of the cycle and to tighten money to deal with the bubbles and frauds preventively. This calls for far more attention to the condition of the technology cycle than has been the practice in the past. The recognition signals that can contribute to understanding of this cycle's progress are the subjects of following chapters.

In particular, once the top of the technology cycle is attained, the appropriate monetary policy is to tighten. In the past the practice has tended to maintain easy money in order to encourage business activity, including investment. Inasmuch as the slate of immediate past investment opportunities has become fully exploited, encouraging business investment seems futile. However, easy money does encourage the pursuit of yield from appreciating tradable assets. The problem has been failure to recognize the top of the technology cycle, and, indeed, such recognition is difficult.

Once the collapse is under way, the Fed's role as a bank regulator becomes as or more important than its role as monetary policy generator. The regulatory role involves making decisions regarding which banks are not so weakened by the crisis as to be salvageable and which banks should be allowed to fail. Successful execution of this role hypothetically will yield a banking system with increased health. The experience since 2007 has suggested that certain rules by which banks are regulated can result in an unrealistically bleak picture of a bank's actual condition. The prime example is the mark-to-market rule. As interpreted immediately prior to the crisis, this rule called for the revaluation of assets to current market value. In this particular crisis, the market for a large volume of housing-related paper went to zero in a current market sense; that is, there were no bids for this paper as the crisis unfolded. While the bids for residential-related paper went to zero, a substantial number of these "toxic" assets still yielded revenues and for this reason, a case for valuing these by means alternative to a dysfunctional current market can be made. This raised a serious question regarding the mark-to-market rule because its interpretation appeared to yield an unrealistically negative view of some banks' true condition. Inasmuch as the purpose of this and other accounting rules is to yield an *accurate* picture of the condition of an enterprise, the rule justifiably came in for questioning.

The role of government regulation or lack thereof in the ongoing crisis is the subject of political debate. There is a saying on Wall St. to the effect that it takes seven years to emplace a new regulation and it takes a clever Wall St. lawyer seven minutes to figure out how to circumvent it. The reality behind this saying is that any attempt by government regulators to microregulate the actions of financial market actors will fail, eventually if not soon. The *constructive* opportunity for regulation lies in the area of improving the transparency of securities markets. This was the objective of mark-to-market rules. Present problems related to these rules exemplify the difficulty of anticipating the impact of the rules under all possible market conditions. Measures to increase transparency may need to be revised with ongoing experience, but that does not compromise the value of increased transparency.

Notes

1. It is fairly unusual that invention results from a scientific discovery; the far more common event is the opposite. Science typically plays the role of explaining why something works after its invention or discovery.

2. The tendency for increases in entrepreneurial activities to appear *en masse* has long been recognized. See, for example, Joseph A. Schumpeter, *The Theory of Economic Development* (Reprint of the 1934 edition) (New Brunswick, NJ: Transactions Publishers, 2005), 223–29. Schumpeter's explanation of this phenomenon did not emphasize the *in masse* appearance of new technologies, however.

3. For an earlier eloquent expression of this idea, see J. R. Hicks, *Value and Capital*, 2nd ed. (Oxford: Oxford University Press, 1965), 297–99.

4. The reader will note that this statement is completely contrary to the ideas that support the U.S. government's present attempts to reignite growth by means of monetary and fiscal policies. Similar policies proved highly ineffective in reviving the U.K. economy in the 1970s and the Japanese economy in the 1990s. A systematic elaboration of the statement appears below.

5. J. R. Hicks, *Value and Capital*, 2nd ed. (Oxford: Oxford University Press, 1965), 298.

6. The reader will note that in an equilibrium state, the rate of return to investment has been driven to the cost of capital. Therefore, the technology-based investment boom grows from an interequilibrium condition. For this reason, economic models based on equilibrium assumptions do not deal with prolonged technology-based upswings very well.

7. The Irr method is but one of the commonly used capital budgeting tools, and is often used in conjunction with the net present value method.

8. It can be argued that the appearance of second-stage innovations of prototype inventions should be predictable because hindsight generally shows that such inventions are logical results of earlier technological development; however, the timing of such an appearance is highly uncertain. Consequently, the emergence of events capable of triggering a technology revolution is an exogenous event for practical purposes.

9. In other words, speculation in tradable assets comes to exceed investment in productive assets as profitable productive opportunities for such investment disappear. This transition is a general diminution of the quality of yields to investment.

10. One product of the pursuit of yield since 2007 was the *auction rate security*. This term refers to a debt instrument (corporate or municipal) with long-run nominal maturity for which the interest rate is regularly reset through a Dutch auction each week or month. These securities offered comparatively high yield and appeared to offer high liquidity and became popular with financial officers of firms of all sizes as places for spare cash. In 2008, most of the auctions failed due to lack of demand with the result that many auction rate securities lost most or all value. Investment banks which had pushed these instruments agreed to repurchase them

at par, but this restitution was far from covering all the losses, and these worthless instruments continue to overhang and inhibit resumption of lending.

11. Hyman P. Minsky, *Stabilizing an Unstable Economy* (New Haven, CT: Yale University Press, 1986), 230–38. In Minsky's view, financial instability grew from traits internal to capitalistic finance. The present argument does not take issue with Minsky's ideas of financial instability, but introduces a cyclically variable impact of the technology cycle as a determinant of the extent of the general economy's vulnerability to the effects of unstable financial markets. The body of economic theory known as the neoclassical synthesis, which dominated economic thought in the post-World War II era, envisioned a high degree of self-righting power in the economy, and saw the threats as exogenous events, such as energy shocks (black swans?). Minsky saw instabilities in the economic structure as more threatening than external shocks. This book proposes what in some ways is the biggest external shock of all: the temporary exhaustion of productive investment opportunities between technology-driven investment booms.

12. In the years immediately prior to 2007, consumption had grown to exceed disposable personal income and was supported by credit and home refinancing.

13. Investment "opportunities" at this stage of the technology cycle are likely to be perceived vestiges of the upswing that are vulnerable to worsened internal cash flows and decreased access to outside financing.

14. A typical response to an economic collapse is to keep interest rates at a very low level in the hopes of stimulating business investment via a low cost of capital. However, if the pool of profitable productive investment opportunities is temporarily fully exploited, this policy encourages speculation in tradable commodities and encourages the development of asset bubbles without doing anything to encourage productive investment.

15. As noted above, asset bubbles are a result of a combination of high liquidity and pursuit of yield.

16. As of the present writing, access to capital remains a major problem for all but the largest and strongest corporations. Hundreds of smaller companies are still experiencing difficulties, and much of the banking sector remains threatened with failure. Other traditional sources of funding for start-up companies have dried up completely. Given the usual role of startups in technology-founded upswings, present conditions are highly unfavorable to the start of an upswing.

2

Stage-Two Innovation and Technology Revolution

There is always something to upset the most careful of human calculations.

—Ihara Saikaku (1642–1693)

As noted above, the second stage of the innovation of a potentially disruptive prototype invention (or group of such inventions or their equivalents) begins when stage-one innovation has brought the invention(s) to a point of practical usability in some market. A second stage of innovation is the setting for a technology revolution, wherein the second stage of innovation is of sufficient power to become a principal refocus for ongoing normal technological development. In a revolution, a massive expansion in the application of the new technology and the obsolescence of older technology induces a surge in general economic growth. Between the first emergence of a novel technology from stage-one innovation and a technology revolution there is a build-up of interest in the new technology. Inasmuch as this build-up is eventually translated into market demand, it is of interest to explore its sources. This is the purpose of this chapter.

A group of prototype inventions or a radical change in regulatory environment can transform a heretofore orderly seeming progression of normal technology into a technology revolution. This involves, among other things, a dramatic change in the sources of capital investment demand: this is illustrated schematically in Figure 2.1. This illustration greatly simplifies a process that can extend over multiple decades, but it captures its essence. The reader will note that the revolution results in a substantial change in the objects of normal technology: the new technology is the foundation of new industry; some preexisting industries become obsolete and disappear; and

Figure 2.1
Investment Demand Sources

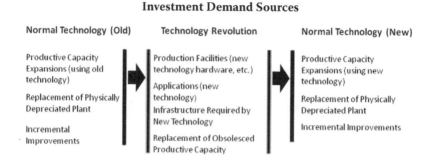

Normal Technology (Old)	Technology Revolution	Normal Technology (New)
Productive Capacity Expansions (using old technology) Replacement of Physically Depreciated Plant Incremental Improvements	Production Facilities (new technology hardware, etc.) Applications (new technology) Infrastructure Required by New Technology Replacement of Obsolesced Productive Capacity	Productive Capacity Expansions (using new technology) Replacement of Physically Depreciated Plant Incremental Improvements

other preexisting industries undergo profound changes as a result of employing the new technology.

Normal technological developments that affect a technology from its first emergence from stage-one innovation continue after the revolutionary impacts of the new technology have subsided (that is, when the new technology has become commonplace); they are ongoing. If the new technology is of power sufficient to bring a large-scale technology revolution, then the normal technological change that grows from it becomes a defining characteristic of ensuing normal technological change. It does not go too far to observe that the transition between today's automobiles and the machines of 1900 falls mostly under the heading of normal technological change. The most important thing to remember about the onset of a technology revolution is that it occurs in response to *a significant decline in the costs of applying the new technology*. Today, there are many threads of technological development which have attracted interest. Indeed, public interest has led to various forms of government encouragement of some of these threads. However, it is questionable that encouragements from government can achieve the conditions of a technology-founded growth cycle without the help of engineering or scientific breakthroughs that decisively diminish costs of application of the technology sufficient to attract rapidly expanding demand across a wide spectrum of markets. One source of falling costs is scale effects associated with a new technology; but scale is not usually the trigger for a technology revolution; that typically requires some engineering breakthrough that is not dependent on production scale.[1]

One way to look at a technology revolution is as a renewal of the slate of improvement possibilities whose exploitation is the function of normal technology. A technology revolution introduces a new slate of incremental improvement possibilities. The ensuing process of normal technology exploits these opportunities. However, the potential for incremental improvement is ultimately limited; eventually it becomes fully exploited. In the early twentieth century, relatively small improvements in the thermal efficiency of the stationary reciprocating steam engine continued to be introduced, but this technology at the time was very mature, and the incremental impacts of such improvements tended to be small. Indeed, decisions to adopt a radically new technology have been influenced by the idea that its improvement potential greatly exceeds that of a mature technology that it replaces.

What Would a Technology Index Look Like?

There is no widely accepted index depicting the state of technological development. Such an index should affect measures of total capital stock; however, the methods by which officially published capital stock measures are compiled do not include any explicit recognition of changes in the productive quality of capital. The state of technology is certainly reflected in the published productivity statistics, but these data also respond to influences other than technology, such as labor input.

Any discussion of a technology index has to be in terms of its effects. However, from these effects, it is possible to understand at least some characteristics of a technology index. Suppose that T is an index of the state of technology. The first difference of T therefore is a measure of the *intensity* of technology change. The "even" school assumes that T_t is a steadily (or even monotonically) changing background to conventional macroeconomic policy measures; in other words, T_t is independent of T_{t-1}, or T_t is essentially like a coin-tossing game. This assumption is equivalent to assuming that the process that generates observed values of T is Gaussian. T_t being Gaussian does not preclude runs of exceptional growth of T, for even an honest coin-tossing game will generate runs of heads or tails from time to time. Given this, can the observed episodes of T growth be attributed to such runs?

T is almost certainly *not* based on a Gaussian process, however, for all students of technology history recognize that current and recent values of T_t are related to earlier values of the index; for this is the implication of the observation that technology change rests on past technology developments. Therefore, T_t is a time series having a complex pattern of long-term serial dependence, a trait that is inconsistent with the conclusion that it is Gaussian.

What can be said regarding the nature of this long-term dependence? There are a number of statistical models for dealing with serial dependence, ranging from simple autocorrelation schemes to complex Markovian systems. However, all of these assume that the total past dependence can be captured in a fixed number of past observations. Inasmuch as the origins of important present technologies can go back far into history, it is doubtful that the full influences of the past can be captured within a fixed number of past observations. For example, the hydraulic press, an important tool in current industrial technology, has antecedents in the seventeenth century. The internal combustion (IC) engine provides an extreme example. The essence of the IC idea is that a fuel is exploded in a confined volume, of which one wall is a movable piston. The hot gasses resulting from the fuel's explosion do work against the movable piston. By that definition, the earliest manifestation has to be the cannon, which appeared in the West in the fourteenth century. In a cannon, the piston has to be replaced manually with each cycle, so that the addition of a mechanical cycling mechanism brings the concept up to the present. For an extensive reinforcement of the complexity of serial dependency in technology, the reader is advised to consult James Burke's *The Pinball Effect*, which traces many trails of causality in technology.[2]

One of the implications of long-term serial dependency is a nonperiodic cyclical pattern in T_t, consistent with interludes of intense technological change (technology revolution) alternating with lengthy intervals of relative technology quiescence (normal technology). A cyclical pattern in T_t underpins the possibility that differences in the intensity of technological change can produce pulses of relatively rapid economic growth interspersed with periods of comparative growth quiescence, and this affects the relations between standard tools of macroeconomic policy and policy results.[3] A stage-two innovation process with generality of applicability sufficient to generate the conditions of a technology revolution—investment boom, widespread obsolescence, improved general productivity, etc.—might possibly be a random event *were it not for the clear relationship of T to previous technology.*

What is the connection between long-term serial dependence and episodes of bunching of innovation activity that this book has characterized as technology revolution? Like many of the mathematical tools which have been emphasized in economics during the last seventy years, the idea here came from a branch of engineering: hydrology. One of the most economically influential natural events from ancient times forward is the annual flooding of the Nile River. However, the magnitude of the annual flood was highly variable, to the point where the agricultural system built around it would produce massive surpluses in some years and famine-threatening shortfalls in grain production in dry years.[4] The dream for eons was to create some way of evening out this cycle. The possibility for controlling the river by means of dams became feasible in the latter part of the nineteenth century. One of the central questions as large dams become technically achievable related to the relation between the height of the dam and the time period of floods to be contained. How high should the dam be to control flooding in a one hundred-year period in relation to a dam capable of controlling twenty-five years of flooding?

At first, engineers treated the annual outflows of the Nile as independent events; in other words, equivalent to coin tossing. Consider a coin-tossing game, in which one player gains $1.00 for each head and loses $1.00 for each tail. Even with a perfectly honest coin, there can be runs of heads or tails, in which the players' win (or loss) position can build up. Not surprisingly, there is a very long-term record of the Nile's annual outflows. If these data are viewed as series of independent events, their distribution is reasonably close to Gaussian. If the annual outflow data reflect an underlying Gaussian process as does the coin-tossing game, there is a rule: the ratio of the difference between a player's maximum accumulated win position and his or her maximum accumulated loss position to the number of tosses increases with the square root of the number of tosses. The hydrologist, taking the number of annual outflow observations in the role of the number of tosses, would conclude that a dam capable of containing a one hundred-year flood should be twice as high as one rated for a twenty-five-year flood.[5]

Subsequent research has revealed that the coin-tossing rule underestimates the height requirement in relation to duration.[6] For the Nile, the ½ power law has been replaced by a power law with exponent .73. The finding of a power law exponent different from ½ was verified by extensive research in flow patterns, not only of the Nile, but of numerous rivers in the world. In each case, the difference between the actual power law exponent and the ½ of the coin-tossing case could be accounted for by the process generating the annual outflow variability.[7] The power exponent in this relationship has been given the name of Hurst exponent, after its discoverer, or H. To the extent that H deviates from the value ½, the deviation constitutes a measure of the long-term dependence in the data series. In the hydrological example, the dam protecting against the one hundred-year flood has to be able to contain the results of a longer run of wet years than the twenty-five-year dam than is implied by the assumption that the outflow data is Gaussian. Serial dependence is like a long-term memory in which events in the (sometimes distant) past can influence current events.[8]

Subsequent research has revealed similar power law relationships in a variety of phenomena that have no relationship to hydrology. Vilfredo Pareto observed a power law in his studies of income distribution.[9] In other cases, the distribution of city sizes appears to follow a power law. In insurance, a power law has been found to govern the odds of any size claim.[10] Examples also include a number of long series of commodity and securities prices. Indeed, an H exponent of greater than ½ has been found to imply the very kind of bunching of growth as would be observed in a T index, which is clearly affected by long-term dependency. H < ½ denotes, by contrast, a tendency for a series to double back on itself frequently in such a way as never to move far from its starting point. The historical evidence of observed technology revolutions suggests that the H exponent for T is in the ½ to 1 range.

Having suggested that the process generating the observations of T_t is not Gaussian, the question remains of what can be said about the statistical characteristics of this process.[11] Consider the first difference of T, or ΔT. A technology revolution may be supposed as an extended run of positive variations. If T were Gaussian, the probability of deviations greater than two standard deviations would be low: less than .04. The historical record since the 1890s includes three technology revolutionary episodes in which ΔT may be supposed to have been large on more than one occasion in each of these. Consequently, the distribution underlying T appears to be one that puts more probability weight into its tails than does the Gaussian. Indeed, the concept of technology revolution hints at the possibility that T not only has sharp period-to-period deviations, but possibly outright discontinuities.

One possible candidate for such a distribution can be found among the family of L-stable distributions.[12] While there is no known general expression for the frequency function of this class of distributions, it is known that they satisfy the following condition. Let f(t) be a distribution function in the L-stable family, and φ(t) its characteristic function.[13] Then φ(t) satisfies the following condition:

$$\log \phi(t) = i\delta t - \gamma |t|^\alpha \left[i\beta(t/|t|) \tan(\alpha\pi/2)\right] \tag{2.1}$$

There are four parameters in expression (2.1): α, β, γ, and δ. The parameter β is the positional parameter, and has the range $-1 < \beta < 1$; if β = 0, the distribution is centered on the y axis; β > 0 implies skewness to the right, and β < 0 implies skewness to the left, or in the negative direction. The expectation of the mean of the distribution is given by δ, and γ is a scale parameter. The parameter α regulates the height of the distribution, or equivalently, controls the weight of probability as between the near neighborhood of the mean and the tails.

While there is no known general expression for the density function of the L-stable family, a special case helps to relate equation (2.1) to common usage. If α = 2, β = 0, δ equals the mean of the distribution, and $\gamma = \sigma^2/2$, where σ is the standard deviation, equation (2.1) is the log of the characteristic function of the normal, or Gaussian distribution. One of the most interesting traits of the L-stable family is that α turns out to be the *inverse* of H, described above. The range 1/2 < H < 1 thus corresponds to the range 2 > α > 1. For α in this range, the data are prone to have the trait of "persistence," which means that an upward movement will be followed by another upward movement with high probability, or to "bunch." This is the characteristic of bunching to be found in the T series (even though virtual), and argues against such bunching as the runs that can occur in a coin-tossing game. The parameter α is thus an index of the degree of long-term serial dependence. No such dependence is indicated by α = 2, and the extent to which α is less than 2 indicates the degree of serial dependence. If α > 2, then the data are prone to offsetting moves, such that an increase in the first difference of T is likely to be followed by an offsetting decrease.

The L-stable family of distributions has the property of *scaling*, which means

$$P(T > u) \approx Cu^{-\alpha}. \tag{2.2}$$

This relation clearly pertains to the positive tail of the distribution, where P is probability and α is called the scaling exponent. C is a proportionality constant. The empirical work related to the L-stable family has largely consisted of estimating α from long data series such as commodity or securities prices. While this is not a straightforward or easy estimation, the overall results have pointed to the conclusion that α < 2 in the economic data tested. This result characterizes the economic data series tested as subject to complex long-term serial dependence, and *not* Gaussian.

The T series presents a different problem: it is an unknown, but serial dependence can be taken as an independently founded given. The hypothesis is that the presence of long-run serial dependence implies α < 2. The question is: if α < 2 implies serial dependence, then does long-term serial dependence imply α < 2? As far as this author knows, there is no mathematical proof—in the classical sense—of the validity of this hypothesis. The argument for its validity is nevertheless powerful. Due to extensive work of Benoit Mandelbrot and others, computer simulations which incorporate long-run dependence in variables generate time series with characteristic "persistence," which means that an upward or downward movement has a high probability of being followed by a similar movement. This trait implies an α value less than 2.[14] As already mentioned, the lower α is below 2—and greater than unity—the more probability weight is to be found in the tails of the distribution. This is an important trait in T, for it suggests a generating process that has a much greater tolerance for large jumps and discontinuities (as in a technology revolution) than the Gaussian distribution affords.

The point of this discussion is that technology revolutions are systematic features in the progress of technology and are not merely occasional runs in a coin-tossing game. The basis for this conclusion is the presence of long-term serial dependence that can be imputed to T, which gives it its characteristic of revolutionary change. This discussion of the statistical characteristics of T does not provide the means to construct such a series. Doing that would require not only the statistical specifications, but also the introduction of causal relationships among the various variables going into T, as well as estimates of the internal correlations that exist among this set of variables.

The implications of probability distributions that impute higher probabilities to large jumps and declines than does the normal clearly go beyond the issue of bunching in technology developments. For example, to assume that the Dow Jones Industrial Average moves according to a normal generating process amounts to assuming that the probability of a very large sudden decline is negligible, but it happened on October 19, 1987, anyway.[15] Much recent evidence has suggested that many financial data series are non-Gaussian, but the Gaussianity assumption persists; the mathematics of it are much better understood than those of distributions with $\alpha < 2$. Moreover, virtually all modern finance theory is based on the Gaussian assumption, as are all the risk models relied on by major financial institutions. This means that these models all underestimate risk. The degree to which massive underestimation of risk played a role in the financial crisis that triggered the Great Recession of 2007–2009 is beyond the scope of this book; yet better understanding in this regard is crucial to future financial market health.

Before leaving the parameters of equation (2.1), the probable value of β deserves mention. It is common to think of technological change as "progress," but there are arguments to the effect that not all of it is. If technology change is viewed as change which results in increased output of goods and services *per capita*, however, it is difficult to escape the proposition that the T series has a strong bias to the positive, or that $\beta = 1$, or at least is close to positive unity.

The Unfolding of Early Stage-two Innovation

Hobbyists

When a technology thread first takes a useful form, one of the first sources of demand is what we would call hobbyists.[16] These include people who are curious about the new object and have enough wealth and leisure time to indulge their curiosity by buying an early offering just to see what they can do with it. This group of people figures conspicuously in the earliest days of both the automobile and the personal computer (PC). In some cases, hobbyists were not simply passive first users of the product, but contributed materially to early stage-two innovation. Charles Kettering, who is credited with the automotive electric starting mechanism, had one of his earliest exposures to the automobile when he undertook to help a neighbor assemble a Packard in his barn; the car had been delivered to the neighbor in kit form.

Hobbyists were a recognizable component in the earliest demand for what became the PC starting in the 1970s. Indeed, the earliest Apple offerings were in the form of circuit boards with no chasses; clearly no one who was not interested in taking on the provision of physical mounting or of programming his/her own applications would be interested. At least some of the standardized software that made the desktop computers of the early-to-mid-1980s useful evolved from the efforts of hobbyists to program their computers. Tandy's early TRS models included such amenities as monitors and keyboards, and early users included individual small investors who found these computers useful in their investment interests.

Wealthy Individuals

This group and that of the hobbyists overlap. Even in the earliest days of the automobile, it was possible for a dedicated hobbyist to indulge an interest in cars. Personal wealth was not necessary for the indulgence, but it helped.[17] The importance of this component of early demand was these people could pay for improvements. As markets for new technologies develop, there is a tendency from a very early stage in the development for improvements to be introduced at the high end and percolate downscale over time. The improvement could migrate downscale because of falling costs: downscale markets meant bigger markets and therefore reduced costs of production. This is a lasting pattern for any technological stream of development. It is very easy to see this over the entire history of the motorcar. For example, the first automaker to incorporate the electric starter was Cadillac (1912 model); the first to adopt hydraulic braking was Deusenberg (1921)[18]; and the first power braking system (vacuum-assisted hydraulic brakes) appeared on the 1928 Pierce-Arrow. All of these features initially applied to luxury models became standard on mass-market vehicles eventually. Early applications at the high end of the market led to accumulated experience in which production costs yielded to scale effects and declined.

It is less easy to associate demand for the latest improvements with the wealthy in the PC era because, in comparison with conditions that existed at the turn of the twentieth century, the later period was characterized by a generally higher level of personal incomes. Not that much time elapsed between the introduction of the 8086 microprocessor and the latest Pentium chip, but there was a substantial demand for computers with processors intermediate between these two. This reflected expansion of appreciation of what might be done with computing combined with at least some frustration as individual and business encountered limits on their existing equipment.

Public Fascination

This is a source of demand for new technology that includes the hobbyist component of demand. Public fascination often took the form of interest in racing, and this was important both with the development of the automobile and with aviation. Auto racing with media coverage dates from before 1900, and aero racing was popularly followed in the 1920s and 1930s. The stress on machines under racing conditions brings out weaknesses, and manufacturers of the equipment responded to demonstrated problems with improved designs. Auto racing took many forms besides racing on one-location courses such as in the Indianapolis 500. There were several well-followed trans-U.S. races before 1910 as well as several round-the-world contests that included cross-Siberian stages. Public fascination with flying supported "barnstorming" of the 1920s as well as the famous trans-oceanic flights of Lindbergh and others. The interwar years, up until the later 1930s, were years of low military budgets, and it was publically supported racing and related activities that brought significant improvements in aero engines and airframes that were put to use in World War II.

Learning Curve

Sometimes it is not entirely obvious what to do with a radically new technology when it is initially introduced to the world. Often, the new technology has an obvious relation to some previous technology, such as the farm tractor. Initially, the tractor was regarded simply as a mechanical mule, and used as a simple replacement for animal power. The tractor became dramatically more efficient with the invention of the power takeoff, but this happened well after the initial invention. The laser, a ubiquitous device in early twenty-first-century life, is unusual in that when it was invented in the late 1950s, apparently no one had a clear idea as to what use to make of it; it was almost as though it was invented just to prove the underlying principle. Its immediate antecedent was the MASER (microwave amplification through stimulated emission radiation), whose design genesis was in the intense research effort on radar that had taken place in World War II. The originators of the MASER saw no theoretical reason why the maser principle could not be applied to visible light (hence LASER). The idea was sufficiently compelling to attract enough corporate support to bring the idea to hardware, but at the time it was invented, it was the subject of the facetious observation that it was a "solution in search of a problem." What ensued was as forceful an example of a learning curve as exists.

Formal Private Research and Development

Once a technology development stream has become established, new companies that have grown up with it will produce ongoing improvements. The historical examples more often than not include the acquisition of rights to a new technology rather than inventing it. A prime example is General Electric's acquisition of rights to the use of tungsten as filament material in incandescent light bulbs. This improvement came well after the initial invention of the incandescent bulb based on less-durable filament material. This example also illustrates much of the emphasis of corporate research and development since: the improvement of existing product lines as compared with the original invention activities that could form the basis of new product lines.

In another example, prototypes of important elements of modern computing were invented in Xerox Palo Alto Research Center (PARC) laboratory. These included the desktop computer, the laser printer, the graphic interface, and the mouse. These elements were developed as responses to time delays that the staff encountered in computing using dumb terminals connected to mainframe computers. Vastly improved versions incorporating the concepts of the PARC hardware have become common today. However, the PARC devices coexisted with the result of hobbyist tinkering mainly because their costs were in excess of those of the early crude Apple and comparable devices.

When a new technology emerges, there are two foci to its development: improvements in the new products themselves and improvements in the techniques for producing them. It is in this latter activity that private industry exerts its most powerful influence on technology development; indeed, it is improved production techniques wherein originate the sharp declines in the costs of applying the new technology necessary to turn a stage-two innovation process into a technology revolution.[19] The early classic example is the dramatic decline in the costs of a variety of industrial products that grew out of the early-twentieth-century development of the electric motor. The recent example is modern electronic computing. Its conceptual history of computing goes back at least to Charles Babbage in the 1830s. The proximate ancestor of modern computing that is often cited was the ENIAC of 1946.[20] From that time forward, various improvements in techniques for data input, storage, and processing led to the early small (desktop and smaller) but powerful computers of today, whose costs have fallen sufficiently low to encourage an enormous expansion of applications in business, industry, and home.

Deliberate Government Policy

Government has historically been a prolific source and/or facilitator of new technology. It was an especially important player in the early days of the republic because it was the only market that was large enough to support production runs large enough to justify design and construction of specialized machinery for the production of parts and subassemblies.[21] The arm of government most involved in this pioneering development was the War Department. It often was the case that the objects purchased in quantity by the government, such as small arms, did not incorporate dramatically advanced technology, but the early defense contracts very much fostered techniques of mass production, especially machine tools.

Many twentieth-century contributions of government were in the form of stage-one development from laboratory to useful device or organization. This is a huge contribution to progress, inasmuch as it absorbs the extreme risk that characterizes stage-one innovation, which the private sector is often reluctant to assume. When the results of government-financed stage-one innovations move into the private sector, there is often a powerful commercial impetus for their stage-two development. The Internet and global positioning system are classic examples of technologies pioneered by government and substantially improved by private industry. Interestingly, many of the technology threads that the government has originated have had some kind of national defense motivation. ENIAC's construction was government-financed and its intended purpose was clearly related to national defense.[22] Other motivations have been related primarily to healthcare and a variety of other problems.

Even though government has produced rich contributions to technology developments, such undertakings have usually not been with the intent of direct inducement of economic growth; the purposes have related to narrow objectives related to national defense, national health promotion, or other goals seemingly unrelated to inducing general growth. Paradoxically, it is difficult to find examples wherein expenditures specifically designed to promote general economic growth actually achieved that objective. The history of direct government efforts to enable economic growth has included the provision of a canal system in the early days of the Republic, various encouragements to railway construction in the nineteenth century, and, in the twentieth century, the provision of a highway infrastructure. These measures *facilitated* growth, but, generally speaking, did not *foment* it. Deliberate attempts to foment growth, such as by depression-era and more recent deficit spending, and by centrally planned efforts to influence the direction of private capital investment, have been tried in a number of other countries,[23] but have not been notably successful. In fact, it is questionable that government expenditures conceived solely to foment growth have induced any net growth at all. The United States has been relatively free of such efforts, and those that have been made were of dubious value. At the present time, the U.S. government has placed a large bet on eventual acceptance of electric automobiles, partly by subsidizing the manufacturing of advanced batteries. This bet presupposes some combination of falling costs and individual willingness to pay a premium for what under present technology will almost certainly be very expensive vehicles. One of this book's cautions is not to dismiss future technological possibilities, but battery technology has been famously resistant to big jumps for over 200 years. As for individual willingness to sacrifice deeply in the interest of what most will perceive as a tiny individual contribution to air purity, a conclusion of unlikelihood seems inescapable.

In one general area, government has exerted considerable influence on the direction of technology development: regulatory policy, both the imposition of regulation and in abrupt change in regulatory regime. Imposition of regulation tends to influence the direction of technology change in affected industries and activities. Profound change in regulatory regime can have an impact similar to a technology revolution: it can be the equivalent of a disruptive prototype invention in its effect on the direction of normal technology change. The effective deregulation of ground freight transportation around 1980 brought a decrease in the cost of moving goods whose effects were spread over the following ten years at least. Environmental regulation has been a powerful influence over the direction of technology development, but it sometimes aims at forcing the wide use of technologies still in stage-one innovation. The positive effects of this kind of forcing are often doubtful.

Another way that government can influence the nature of normal technological change is by taxation policy. Over the first three-quarters of the twentieth century, European government concerns over the security of gasoline supply led to a taxation regime that encouraged the use of gasoline engines with relatively small displacement and of diesel engines as automotive power. By contrast, North American tax policy encouraged the development of relatively large engine displacements and little development of automotive diesels.

The Unfolding of a Technology Revolution

When important changes in basic technology are introduced, it is natural to assume that it is economically worthwhile for all affected firms to adopt the new technology quickly. In the technologically normal economy, this is the idea behind the treatment of obsolescence as simple augmenter of capital wastage. Increments of technological change are small enough that they can be recognized and incorporated into investment decisions quickly. In the technologically revolutionary economy, instant recognition of the implications of a new technology typically does not happen for a number of reasons. Of these, perhaps the most important is that technology revolutions sometimes open inauspiciously. It is only after a sometimes prolonged learning experience that the full potential and force of the new technology becomes perceived. Moreover, the infrastructure that the new technology requires cannot generally be provided instantaneously. Other delays to immediate scrappage include various kinds of attractive economics of retaining older capital in service, various problems in the establishment of new companies, regulation, and union activity.

A technology revolution does not break upon the world all at once. It starts with the kind of thinking that is characteristic of the technologically normal economy: the earliest ideas typically consist of simple replacement of some component of extant productive systems with something better. When a revolutionary new technology first emerges in a commercially viable form, the field of firms that can quickly adopt it is limited to those in the industry for which the earliest commercial manifestation of the new technology is clearly applicable.

Consider a simple example: the corporate planning function. These groups carry out a variety of functions, including economic analysis of capital investment proposals, mergers and acquisitions, longer-run economic prospects, accounting activities related to planning, and other functions. In the 1970s, the common analytical tool was the spreadsheet aided by the desk or hand calculator. At that time, a spreadsheet was literally a large sheet of paper with ruled lines and columns. An assignment would come to an analyst in the form of a broad outline. This would bring a response in the form of an analysis set out on a spreadsheet. This spreadsheet would often trigger questions from management, such as what would happen to the result if this or that assumption were changed, whereupon the analyst would turn to pencil, eraser, and calculator and revise the analysis, cell by cell. If this sounds laborious, it was. If it happened to be late afternoon and the topic was of high interest to senior managers, then the analyst could look forward to a very late supper.

The 1980s brought the electronic spreadsheet, one of the several early applications that enhanced business interest in the PC. The electronic spreadsheet probably saved some analyst time in the initial setting up of a problem, but the real payoff was that it made responding to request for modification easy. When initially introduced, the electronic spreadsheet had the effect of increasing the efficiency of the existing analyst staff. After this initial stage, however, it became apparent to managements that the electronic spreadsheet had created possibilities beyond making the present staff more productive, for an analyst with a dedicated PC, in addition to having an electronic spreadsheet, had a word processor for the composition of interoffice memoranda. These changeovers thus diminished not only the need for secretarial support, but for analyst time (at least relative to a given work load). Did the introduction of the PC create obsolescence of equipment? The answer has to be yes, at least indirectly. What it unquestionably did was save large amounts of analyst time.[24]

Absorption of the new technology by the wider economy is part and parcel of a sometimes prolonged learning curve in which the full potential of the new technology comes to be realized widely. One of the first commercial applications of the transistor was the transistor radio, a simple replacement of the vacuum tube in an otherwise widely used product. One has only to reflect on how the transistor, as a component of integrated circuits, has become an essential part of the technology landscape of the early twenty-first century. The transistor was invented in the late 1940s and continues to be utilized in devices which have been since developed and widely distributed. Its use has led to the retirement of countless electronic and mechanical devices since its earliest applications.

Clearly the obsolescence in response to the transistor has not always been a quick process, but has reflected expanding realization of the possibilities of transistorized circuitry and the declining costs of such circuitry. It has been anything but a one-time process. In the 1960s, the transistor was widely applied, in the form of amplifier circuits, to playing recorded music, for which the then-popular medium was the LP record. That recording medium became obsolete with the introduction of magnetic tape cartridges (eight-track followed by cassettes) and, later with compact disks. Thus there was a chain of obsolescences, consisting of vacuum tubes, record turntables, eight-track players, and cassette players.[25]

One of the more dramatic examples of obsolescence during the IT revolution comes from the impact of fiber optic cable application on existing communication facilities, of which most were based on copper conductor. In long-distance facilities, fiber optic cable required considerably fewer repeater stations than copper conductors, thereby rendering the equipment of many repeater stations obsolete. In an even more compelling example involving communications lines in very densely developed places, such as Manhattan, copper communications lines were emplaced in tunnels drilled in rock which were physically full and had no room for fiber cables alongside existing lines. In these situations, installation of fiber necessitated removal of the copper, which became scrap copper. For at least a while during the 1980s, this kind of replacement made AT&T a significant factor in the scrap copper market and affected the price of copper in general.

Computer hardware technology improved so rapidly after 1990 that falling costs of computing made application of the technology to highly complex business problems economic. The hardware technology has moved so fast that obsolescence and scrappage of older computers accelerated sharply. In an extreme current example, the computer, in its role as Internet access tool is a major communications facility, now threatens obsolescence of much of the capital stock of the U.S. Postal Service. The Post Office monopoly was created by public law, and the example demonstrates how difficult it is to find an example of a truly stable monopoly, even one enabled by government sanction.

Elements of the Learning Curve: Infrastructure Provision

The infrastructure necessary of the full exploitation of a radical new technology typically does not exist when the technology first emerges as commercially viable in some application. An example demonstrates this process as it worked in the IT revolution. In 1980, the oil and gas industries faced the problem of how to utilize huge volumes of exploratory data to improve their chances for avoiding costly dry holes, in the case of oil and gas. The data came from multiple boreholes and seismic exploration results. The huge number of data points resulted from the sheer geometry of the problem: a relatively complete picture of even a small three-dimensional section of the earth's crust requires very large numbers of observations.

At that time, the most powerful tool for this analysis was the supercomputer. However, the ability to manipulate the dataset was sharply constrained because of its size. Today, computing successfully produces much more finely detailed pictures of the subsurface than was possible in 1980. Present capability includes the ability to "see" through salt layers thousands of feet thick that posed what seemed like an insurmountable problem for the old supercomputers. Part of the difference is, of course, the availability of far more powerful computers which have been developed according to Moore's law. The supercomputer has been replaced by hundreds of PC-sized units. These work in a parallel processing mode, in which the analysis problem is broken down into many parts which can be analyzed separately and brought into the overall problem, somewhat in the manner of the coordinated production traits of the pioneer mass production lines.

Greatly enhanced raw computing power is but part of the change, however. Achievement of today's analytical power required development of the software required to get all those PCs to work in coordinated concert. This illustrates the general nature of the infrastructure development that enables full utilization of the hardware products of the IT revolution. This has made it possible to produce in a few hours graphic results in detail not attainable by the supercomputers, which produced reports of limited value by today's standards. This overall process was driven by the rising costs of drilling wells, especially offshore in deep water; the result has been a significant improvement in success rate.

A friend of the author builds and races formula-one cars. His shop operation includes a computer-controlled machining center, which replaced a group of individual manual lathes, milling machines, and other machining equipments. His experience has been that the new machining center does in three hours work that previously required thirty hours. This is yet another example of the high potential capacity of computing plus the infrastructure of the software and machinery necessary to control many separate machining operations.

The classic example is the automobile, which from the time of its invention in the mid-1880s until about 1900 was little more than a hobby of the well-to-do. The transition of the automobile from hobby to the mainstay of overland transportation was well under way by 1910, and was supported by the development of a passable road net beyond the limits of the larger urban areas, the ongoing development of suitable fuels, and of a refueling and repair infrastructure. The development of suitable fuels was especially important as automobile use expanded, both from the standpoint of the automobile user and of the general economy. Maximizing the gasoline yield from the barrel of crude oil became the primary driver of the development of the petroleum refining industry in the United States, and remains so to this day. This meant not only an increase in the yield of gasoline from crude oil, but also an increase in the octane rating of gasoline. After 1910, automobile factory sales expanded rapidly until 1923, and resulting obsolescence in such areas as public transportation and older crude oil processing facilities was correspondingly rapid.[26]

The obsolescence impact of development of infrastructure that is called for by radical new technology is highly complex. The early building of a network of intercity highways replaced, in many cases, dirt and gravel-surface roads, which themselves became obsolete, but in most cases, the *routes* taken by the newer highway net were the same. A more compelling example of obsolescence resulting from the developing of automotive infrastructure comes from the petroleum refining industry. In spite of the impression that concern with fuel efficiency is solely a recent phenomenon, thermal efficiency of all heat engines has been a driving design concern since heat engines first emerged as commercially viable power units in the eighteenth century. Early designers of IC engines realized that thermal efficiency could be related to one parameter: compression ratio. The higher is the compression ratio, the more thermally efficient is the unit.

Elements of the Learning Curve: Retention of Obsolete Capital in Service

Even when base-load productive capacity is equipped with new equipment that is of substantially different design from that of incumbent equipment, it is a common practice to retain some of the older plant in service for light-duty assignments or as overload capacity. The reason is that the economic analysis justifying the purchase of late-design capital assumes full-capacity operation—24/7 operation in some continuous process industries. When a business experiences cyclically variable demand, it typically does not pay it to acquire relatively costly late-design plant adequate to cover demand peaks.[27] This means that continued operation of some plant that is in some sense "obsolete" can offer attractive economics.

Several examples of retention of "obsolete" equipment in service come to mind. A number of railroads kept steam locomotives in service well after the companies had committed to complete dieselization during the 1950s.[28] The steam engines were employed in such services as seasonal grain movements and in light-duty branch-line services.[29] Dieselization of the railroads took place approximately between 1935 and 1957, and the process of replacement provides an excellent example of prolongation of the obsolescence–scrappage relationship. The diesel cycle engine found its earliest commercial applications after 1897 as stationary engines and in marine applications. There was a steady improvement in the diesel's power density such that by the early 1930s, it was applied experimentally to railroad prime movers. By 1957, few steam locomotives remained in service. Thus a technology that first appeared in 1897 was causing significant scrappage into the 1950s![30]

If obsolete capital is held in service as peaking capacity, its economic life is extended, for economic life is reckoned in working units such as hours or miles. Keeping older plant for intermittent uses delays the retirement of the plant in question, typically until such plant comes due for a major scheduled overhaul. At a reduced rate of usage, the older capital plant may remain in service for an extended number of years. In any event, it is not correct to assume that all equipment recognized as obsolete is scrapped in the year in which it is first recognized as obsolete.

How long can obsolescent capacity be kept in service, even as overload capacity? The limitation on the life of such capacity often takes the form of sharply increased costs to keep it in use for any service. If an industry is in transition to a new technology, most of its investable dollars are being spent on the new processes; there is a reluctance to spend heavily on capacity regarded as obsolete. Any industrial system depends on outside suppliers of critical components and services. As productive capacity based on a technology being displaced diminishes, the firms supplying the obsolete plant either go out of business or diversify into other lines of business. Users of the obsolescent technology are forced to use alternative sources, such as used parts, which become increasingly costly. Perhaps of even greater importance, the population of workers capable of servicing the displaced technology tends to disappear through aging and lack of replacement. When Amtrak took over the railway passenger services in 1971, it attempted to operate with rolling stock that used steam heat and cooling. Use of steam as a refrigerant developed in the 1920s and made sense because the train was powered by machines that included large-capacity steam generators. In Amtrak's early years, these steam-based systems would often fail because competent maintenance labor was rapidly disappearing due to retirements and failure to train replacements; the problem was not solved until the entire steam-based heating and cooling technology was replaced with electric systems.

Elements of the Learning Curve: Regulatory Issues

There are unusual instances in which government regulation prevents early replacement and scrappage of obsolete equipment. These occur in wartime situations, but none have occurred since World War II. In that war, a Federal agency, the War Production Board (WPB), had veto power over industrial investment proposals, and tended to veto any investment that involved equipment that was not deemed proven technology, that would "unnecessarily" divert resources from current production, or anything that struck the board as novel. The object of the WPB was to assure maximum war production by not allowing any diversion of effort or material to experimentation that might be necessary to bring novel designs into reliable production.[31]

There is at least one other special case in which regulation affects the pace of scrappage of obsolete capital plant. Under Interstate Commerce Commission (ICC) regulation, railroads faced considerable difficulty in withdrawing from unprofitable businesses. These situations included freight services over certain branch lines, and passenger services. Both of these business lines tended to migrate to highway carriage over the years, but under ICC procedures, local interests could protest an announced abandonment, and the ICC sometimes required continuation of the services in response to these protests. Protesters often were shippers who wanted the railroad present as a brake on highway shipping rates. Railroad deregulation in 1980 resulted in the scrappage of considerable trackage and rolling stock previously employed in these services.

Elements of the Learning Curve: Union Activity

There have been instances wherein labor unions have effectively opposed the application of modernization, but these have for the most part turned out to be delaying actions. There have been many cases in which unions have recognized the inevitability of modernization and worked with managements to minimize the impacts on their members and to protect members' interests as enterprise stakeholders. In other cases, unions have sought to blunt the impact of modernization by sometimes successful opposition to managerial flexibility in worker assignment that would maximize the savings that the modernizations could provide. This sort of union contract effect in some of the older large industries, such as steel and automobiles has tended to break down as these industries encountered serious foreign competition in the later postwar period.[32]

Union activities seemingly unrelated to applications of new technology can affect business decisions regarding capital replacement and scrappage. The United Mine workers under the leadership of John L. Lewis struck the coal industry in 1949 in an action that was not settled until well into 1950. At that time, railroad fuel was the second-largest market for coal after electric power production. One of this strike's effects was to induce a number of railroad companies which previously intended to continue operation of steam power in a number of situations in which this was an economically attractive strategy, to hasten replacement of all steam power. The effect was to accelerate the elimination of the railroad fuel market for coal.

This comes out with reading in the railroad industry press just prior to the walkout; the experience of one railroad is illustrative.[33] The Baltimore & Ohio railway operated a number of steam engines of recent design in 1949, and the company had an active program for bringing a number of its older engines up to the mechanical standard of its newest steam power. This steam policy made economic sense because it provided tractive effort for a smaller investment than the alternative, the acquisition of new diesel locomotives, and because coal fuel was perceived as relatively cheap in its operating territory. The strike upset the plans for utilizing its modern steam power by calling into question the reliability of the fuel supply for these engines. As a result, the modernization investments for the older steam power were canceled, thus shortening their economic lives and hastening the end of the company's use of coal fuel.

Elements of the Learning Curve: Managerial Resistance

It is not uncommon for a revolutionary new technology to be perceived as a threat to a firm that is an established seller in a market.

A firm that has a substantial share of a market may well persuade itself that the new technology does not constitute a serious threat to its established position, and that the cost of developing the new technology will not result in a net gain to its total business. Eastman Kodak's most important product in the 1980s was X-ray plates based on silver halide technology. At about that time, the company's research and development laboratories pioneered in the development of digital photography. It is not difficult to visualize why the company failed to develop digital, for at the time the digital image quality was inferior to silver halide images and it was comfortable to suppose that digital images could never attain sufficient quality to challenge the established business. When the image quality gap was rapidly closed in the ensuing years, Kodak entered the digital field, but by then it had sacrificed any first-arriver advantages it might have enjoyed. Xerox failed to commercialize the PC that its own PARC laboratories had developed. The reason seems to have been senior management's failure to envision a substantial market for PCs and their peripheral equipment.[34] Polaroid made a belated attempt to enter the digital imaging market, but this effort was too feeble to forestall the company's bankruptcy in 2001.

Technology Revolutions: Pre-1930

The first technological revolution of the twentieth century took place roughly between 1905 and 1925. It grew out of two broad threads of new technology: the automobile and electricity. While it is customary to speak of an "automobile industry," the impact of the automobile was far wider than can usually be attributed to a single industry. The roughly contemporaneous emergence of electricity as a fundamental tool in industrial production as well as lighting for both residences and businesses would have amounted to a technology revolution in its own right. As it happened, the combined macroeconomic effects of the two developments underpinned a general economic upsurge that lasted from before 1900 until the mid-1920s. Indeed, the two major technology threads were intertwined, as electric power made possible the precisely coordinated production lines that were the bases of mass-produced low-cost automobiles and other new products.

The revolution saw the evolution of the automobile from hobbyists' object to the basic means of overland passenger transportation in the industrialized world as well as in the United States. While there were antecedents going back to the early nineteenth century, invention of the modern automobile is often credited to Karl Benz, who combined an Otto-cycle engine with a carriage in 1885. By the turn of the twentieth century, there was a substantial, but still relatively small, market for motor cars. A number of new companies had emerged to manufacture autos. It does not exaggerate to say that there was a large and growing public interest in the early cars. Hobbyists were very much a part of this early-day demand, and among hobbyists there were "tinkerers," some of whom contributed materially to the technical development of the automobile. Auto racing drew much interest from very early in the new industry's history.

Prior to 1900, the primitive auto industry adopted improvements in enabling technology and lowered the cost of auto ownership within the range of wealthier individuals. At some time between 1905 and 1910, there began a dramatic production scale-based decline in the cost of owning an automobile, a precondition for the revolution in which the automobile developed into the fundamental means for overland transportation of people. This cost decline resulted proximately from mass production methods, a development to which many contributed, but the process came to be dominated and personified by Henry Ford. Ford was a self-educated engineer of great talent. The production facility of the early Ford Motor Company was little more than a large barn in which teams of workers built cars in bays. Ford's Model T appeared in 1907, and represented enough improvement of Benz' invention that it could aspire to be a popularly demanded vehicle, as it indeed became. It initially sold for the asking price of $875.00.

The opening of Ford's Highland Park, MI, plant in 1913 was a historical milestone in the progress of low-cost mass production. It was a true mass production facility, in which workers stood in one place and performed one operation on unfinished autos as they moved by, dragged by a chain-like moving platform. The process included a main production line and subsidiary lines which brought subassemblies to the main line where they were incorporated in the unfinished vehicles. Workers, according to Ford's intent, did not have to move their feet.[35] The system was enormously productive and was widely emulated eventually. Ford's pricing policy sought the mass market by reducing the price of the Model T as it production costs declined, and by 1920, the Model T held about 60 percent of the U.S. auto market. The Model T continues to hold the record as the most reproduced automobile in all history.

The technology revolution took place roughly between 1905 and about 1925. The capital investment boom that it fostered grew not only of the expanded demand for steel, glass, and other elements of the automobile itself, but also of demand for paved roads and highways, and demand for industrial facilities in general designed around the distributed power enabled by the electric motor. It was essentially over by the mid-1920s when markets for its main products became temporarily mature.[36] Its genesis was in the dramatic decline in the costs, not only of automobiles to individuals, but also in the falling costs of electric power to industry and households.

Its exhaustion made the general economy vulnerable to a financial crisis, which occurred in 1929. The great depression that followed bore the earmarks of an investment failure that reflected the full exploitation of the slate of investment opportunities that fomented the preceding technology revolution. This period of technology revolution was characterized by incremental improvements in the automobile and in electrically driven industrial plants of the kind associated with normal technological change. The difference that the revolution made was that the normal technology threads that preceded the revolution changed radically as possibilities growing from electrification and automotive developments became dominant threads of normal technology. Take automotive brakes, for example. Modern hydraulically controlled drum-enclosed brakes began to appear on luxury cars in 1921, rather late in the automotive revolution.[37] Many other mechanical improvements appeared throughout the years following, including the years of the great depression.

This onward march of normal technological change probably contributed to the severity and longevity of the depression. Even though the growth in automobile factory sales had ceased after 1923, the automakers continued to invest heavily in production facilities. In the face of falling market share, for example, Ford invested heavily in new production facilities for its replacement to the Model T, whose production ceased in 1927. Other improvements during the 1920s called for lesser investments. Many of these investment threads were curtailed or postponed when demand softened in 1930 and later. One can wonder why heavy investment continued in the late 1920s amid signs that the market was mature, and one can only conjecture that managements steeped in the experience of ongoing growth were psychologically unable to cope with the idea of a massive reversal of this experience.

The Information Technology Revolution

The IT revolution was based on the integrated circuit, one of the most important prototype inventions of the twentieth century, along with fiber optics, and the laser. The charge-coupled device (CCD), which led to a revolution in imaging technology during the 1990s, had many of the characteristics of prototype invention. It underpinned the healthy general economic growth of the U.S. economy and the economies of other industrialized nations starting in the 1980s and lasting until about 2005. It was not the only technology revolution after that of pre-1925, and it differs from the earlier experience in many details. Yet, an examination of the IT revolution reveals some broad traits in common with the early twentieth-century experience. Four stand out: (1) it was preceded by a sometimes very long history of the normal technology developments of its physical elements; (2) it was immediately preceded by a dramatic decline in the costs of applying its technology; (3) its initial genesis was fostered by hobbyists and wider public interest; and (4) the technology was commercialized largely by new companies founded for the explicit purpose of exploiting the new technology.

From ENIAC, the electronic computer evolved fairly rapidly into a business and industry tool that was adapted to an expanding variety of business problems. Early examples included payroll management. Major technological events in the development of computing included the introduction of solid-state devices (transistors) in place of vacuum tubes, and, later, integrated circuits. By the early 1980s, a number of the early components of modern computing, such as increasingly powerful central processing circuitry and laser printers had begun to appear in forms clearly recognizable as direct antecedents of today's technology. There began a period of falling costs and increasing power of computing that led to business' ability to harness computers to an expanded array of business problems of increasing complexity. There was a parallel improvement in the power of software in the form of operating systems, and business applications such as word processors and electronic spreadsheets. Moreover, as integrated circuitry became increasingly powerful, its application spread well beyond computing as narrowly defined into various consumer and business devices, such as cellular telephones.

It is easy to suppose that the principal impact on the general economy of these developments occurred after 1990, but there is considerable evidence that the impact was present before then. One of the more telling evidences was the disparate effects of expansionary monetary policies as between the later 1970s and the mid-1980s: worsening inflation in the earlier period and diminishing inflation in the later period. In the 1980s, there was an expanding field of productive investment opportunities in the gathering IT revolution; this technology support was mostly absent in the previous decade. The economy's ability to absorb investable liquidity was much better in the 1980s. The most compelling evidence of the investment boom's having run its course by the mid-2000s is circumstantial: many of the markets whose growth had fostered the boom had become mature. Technologies that were revolutionary in the early 1990s had become commonplace by 2005 and had no more potential for inducing capital investment than is characteristic of normal technological change.

The IT revolution and the investment boom that it engendered were interrupted by several recessions triggered by financial market events, of which the worst occurred in 2000–2001. This recession had two (not unrelated) triggers: a stock valuation bubble centered in the stocks of high-technology companies, and the recognition of a huge overinvestment in telecommunications facilities, especially fiber optic cables. Apart from the telecommunications sector, growth in other areas of high-tech hesitated, but did not go negative. The stock market decline was centered in telecommunications-related stocks, but affected other stock valuations. Recovery was not especially prolonged, at least by 2007–2009 standards, indicating that the IT-generated investment boom still had some life to it. However, the event did indicate that at least part of the markets on which the IT revolution had fed was mature.

The Prosperity Following World War II

The general prosperity that characterized the U.S. economy from the end of the war and lasted into the 1960s had technological foundations, but conditions of these years were influenced by several elements not present in the pre-1930 or the IT revolution. The most important of these was the war itself, which had at least three very important after-effects. The first was that it effectively eliminated foreign competition to U.S. industry. The second was that wartime rationing created substantial "forced" savings among American consumers, and this fund financed a consumer expenditure boom immediately following the elimination of rationing and prevented the economy from falling back into the depressed economic conditions that preceded the war. The third was that the war delayed the development of several consumer-oriented technologies that had emerged in the Depression years.[38]

The relative absence of foreign competition in the early postwar decades meant high profitability for a number of established industries such as automobiles and steel. Industrial labor unions that had emerged in the 1930s successfully sought larger shares of the economic rents implied in the high profits of the era, and this had the macroeconomic effect of lifting the incomes of many factory workers into the middle-class range, with salutary effects on overall consumer spending. Reviving foreign competition in these and other heavy industries created problems for the industries and their labor forces by eliminating some of the economic rents that had been the bases for shared labor-industry prosperity prior to 1970.[39]

Once these unusual conditions are taken into account, the technology foundations for a general prosperity become discernable. One of the most important of the new technology threads was the growth of commercial aviation. Aero engines and airframes had improved dramatically in the interwar period. Government defense spending played a minor role in this development before the second half of the 1930s; much of the support was derived from massive public interest in aviation. Aero racing was a popularly followed sport, and, as in the history of automobile, failures under racing conditions led to mechanical improvements. This not only provided a technological foundation for high-performance warplanes, but some of the aircraft that were successful in the war were adapted to the needs of civilian air transport.[40] Military jet propulsion emerged in the very closing months of the war, was developed for military use during the 1950s, and moved into the private sector in the form of the Boeing 707. The macroeconomic importance of expanded commercial aviation has to be reckoned in terms of travelers' time. In one homely example, the emergence of jet travel enabled the expansion of major league baseball from the northeast and midwest (the range enabled by rail travel) to the entire country.

A second major technology thread was television. TV had emerged in crude form in the late 1930s but its consumer development was put on hold during the war. However, home sets proliferated rapidly in 1948–1949. TV pictures were not always well resolved, and picture tubes were small. Early proliferation of sets and the development of broadcasting can be attributed heavily to the public interest factor.

As in air travel, the macroeconomic effect can be traced to the value of time. For example, conventional photographic reportage in the Korean War could get an image into *Life* in a week. A TV image could make it to newscasts almost on the day created. Part of the basis for the general prosperity was the investment in TV manufacturing facilities. This was in the hands of a number of relatively new companies that had roots in radio broadcasting in the 1930s.

Several other technology revolutions were in progress during the early postwar decades. These included dieselization of railroads and motor freight, as noted earlier. Both of these developments increased the efficiency of moving goods overland, but in different ways. On the railroads, the replacement of steam locomotives required replacement of the substantial infrastructure needed to operate a fleet of steam locomotives, and the diesel infrastructure was far less labor intensive than that for steam. The motor freight industry realized substantial cost savings from the transition due to the increased thermal efficiency that dieselization offered. In both of these industries, full realization of the possibilities of the new technologies had to await deregulation, which took place in 1980; but the economies that were realizable in the 1950s were substantial.

The markets that supported the products of technology in the postwar decades were mature by 1970. In the decade that followed, there developed serious foreign competition in automobiles and basic industrial goods and a tendency for important parts of the consumer electronics industry to migrate offshore. There was no spectacular financial failure that triggered a descent into hard times comparable to those of 1929 or 2007–2008 at the outset of the decade, but there also was little support for growth in the form of a technology revolution. Efforts to induce growth with monetary policy produced an inflation problem but little growth (this condition was often referred to as "stagflation," or the simultaneous occurrence of growth stagnation and inflation). Recessions that did occur in 1973–1975 and 1981 were severe by then postwar standards.[41] Such technological change that did occur in the 1970s was strictly of the "normal" character.

Technology Revolutions and Countercyclical Policy

A point stated earlier can be developed here: the presence or absence of a technology revolution in progress makes a large difference in how successful standard tools of public economic policy, especially monetary policy, will be. Financial crises are not prevented by technology

Figure 2.2
U.S. Real GDP: Annual Percent Change

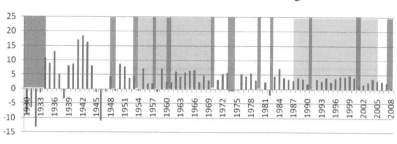

Note: Lighter-shaded intervals are approximate periods of technology-driven investment booms. Heavier-shaded intervals are recessions. Recession data are from the National Bureau of Economic Research.

revolutions (Figure 2.2). The hypothesis here is that recessions will be milder and more easily handled by monetary and financial measures if they are implemented with a background of an ongoing investment boom based on a technology revolution. Using the approximate timings of technology-based investment booms since 1929, the peak-to-trough average duration for recessions occurring during investment booms is 6.8 months. For other recessions, the comparable average is seventeen months (based on National Bureau of Economic Research [NBER] data). The suggestion here is that an active investment boom is a substantial help in the success of monetary policy, which has been the principal countercyclical policy tool in the post-World War II era.

This point can be expanded by means of a review of four financial crises in comparably recent history: the savings and loan crisis of the late 1980s; the "Black Monday" crisis of October 1987; the failure of Long-Term Capital Management (LCTM) in the late 1990s; and the collapse of residential and commercial construction following 2007. In each of these examples there was an intervention by the Federal Reserve System of some kind for the stated purpose of warding off a major recession or worse. Of these examples, the first three occurred during the period herein identified as that of the IT revolution and its resultant investment boom.

LCTM was a hedge fund founded in 1994 and operated on the premise that valuation anomalies in various securities markets could be detected by means of modern quantitative methods. The fund operated with high leverage.[42] It was highly profitable for the first several years of its operation, but lost heavily in 1998. This produced a liquidity problem, and monetary authorities became aware that LCTM was in serious danger of failure. Moreover, inasmuch as LCTM had practically all the major Wall St. financial houses as counterparties, it was feared that its failure could bring down the national and world financial structure. When LCTM failed to raise capital on its own, it was "bailed out" by substantial loans from threatened counterparties organized by the Federal Reserve Bank of New York. The fund survived until 2000 when it was liquidated. All loans were repaid, and no recession ensued.

The savings and loan crisis refers to the failure of over 700 savings & loan companies in the late 1980s. There were a number of causes, but a detailed investigation of these is beyond the scope of this book. The industry was bailed out via a combination of measures whose effect was to visit costs on taxpayers. Between 1986 and 1991 the number of new housing units constructed fell substantially, and the affair is suspected of having contributed to the brief recession of 1990–1991. The recession was countered by an injection of liquidity into the economy by the Federal Reserve which is credited with limiting the damage of the recession.

The "Black Monday" crisis refers to a stock market crash that occurred on October 19, 1987. The crash followed a sequence of similar crashes that started in the Far East and circled the globe. The causes of these events are still being debated, and it is not the purpose here to resolve these debates. What is of interest is the Federal Reserve's reaction was to inject massive amounts of liquidity into the economy, and no recession ensued.

These three examples occurred during the buildup of the IT revolution when the economy had ample opportunities for absorbing liquidity productively. The response to the housing crisis of post-2007 provides sharp contrast. The Federal Reserve appears to have deployed the most powerful weapons in its arsenal, but the recession occurred anyway, and while it is over by NBER definition,[43] the recovery has been weak. The suggestion here is that the economy lacks much of the ability to absorb liquidity productively that it had in the 1990s and late 1980s.

Does this kind of comparison hold in the decade prior to 1980? As noted above, the 1970s economy is remembered as the decade of stagflation, in which the U.S. economy showed a tendency toward inflation. The recession of 1973–1975 appeared to arrest inflation but only temporarily, for in the years following 1975, inflation gave all the appearances of becoming increasingly higher. Expansive monetary policy during this period was probably the opposite of what was needed to arrest what appeared to be a long-term inflationary tendency, but it also utterly failed to foment solid economic growth. In terms of the technology cycle, the 1970s decade was an interval between the technology push that had enabled low inflation growth in the 1960s and what became the IT revolution of the 1990s. The early part of the decade was a time of unfortunate experimentation with price controls which ultimately proved ineffective at countering the recognized inflationary tendency. It should be pointed out that there was not a strong technology underpinning for the economy in which some of the ample liquidity could have been absorbed productively; instead it underpinned inflation.

Notes

1.   Anticipated cost consequences of large-scale production are often involved in decisions to invest in volume production facilities. Such anticipation has occurred in a number of cases in the high-technology industries. However, such decisions have usually been preceded by cost-reducing breakthroughs which have brought costs down to the point from which scale economies can realistically be anticipated.

2.   James Burke, *The Pinball Effect* (New York: Little Brown, 1996).

3.   "It is a peculiar property of most long-memory processes that *seeming* patterns arise and fall, appear and disappear. They could vanish at any instant. They cannot be predicted." Benoit Mandelbrot and Richard L. Hudson, *The (Mis)behavior of Markets: A Fractal View of Risk, Ruin, and Reward* (New York: Basic Books, 2004), 189.

4.   For example, see *Genesis: 41.*

5.   The longer time period of control of the higher dam being four times the shorter, application of the square root rule suggests the square root of four as the planning height, or twice the height of the twenty-five-year dam.

6.   In the coin-tossing example, a plot of the difference between the maximum win position and the maximum loss position against the total number of tosses on log–log paper yields a straight line with slope ½. The hydrologists found that their plots of the ratios of maximum flow difference versus number of years also yielded the straight line indicative of a power law, but having slope greater than ½. An H value of ½ (the coin-tossing case) corresponds to the condition that each event is independent of all other events. An H value greater than ½ indicates the presence of serial dependence.

7.   In the case of the Nile, the annual flooding comes principally from the variation in the flow rate of a major tributary, the Blue Nile. The Blue Nile arises in the Ethiopian Highlands and is subject to a wet–dry alternation that is usual in tropical climates. The *cyclic* variation reflects a pattern of

runs of wet years alternating with runs of dry years. It is this cyclic pattern for which the dam designers have to allow.

8. Mandelbrot and Hudson, *(Mis)behavior of Markets*, 173ff.

9. Pareto is remembered for his work in general economic equilibrium; his work on income distribution has largely been forgotten.

10. Mandelbrot and Hudson, *(Mis)behavior of Markets*, 152–57.

11. A stationary time series is one whose underlying generating process can be described by one distribution function, such as the normal in some cases.

12. The L-stable distributions have the useful property that a sum or other linear combination of L-stable variables is also L-stable. This property makes these distributions appropriate for characterizing economic data which are themselves summations. Income, a summation of a number of incomes from different sources is an example. The mathematics of the L-stable family goes back to Paul Levy in the 1920s. The present summary is drawn from Mandelbrot (1997, 446–47) and Mandelbrot and Hudson (2004, 173ff).

13. If the characteristic function of a distribution is known, then the distribution exists whether or not its explicit form is known.

14. Mandelbrot is known as the father of fractal geometry. A fractal is ultimately a limit of a process in which a simple relationship of geometric form is subjected to many iterations. Mandelbrot has developed the technique of generating time series that strongly resemble actual price histories from iterations of simple geometric forms. He freely terms these results as "cartoons" and so far is not able to reproduce actual historical data fractally. This technique is a central part of his *Fractals and Scaling in Finance* (New York: Springer, 1977). See also Mandelbrot and Hudson, *(Mis)behavior of Markets*, 194.

15. This was the infamous "Black Monday" in which the Monday opening value of the Dow Jones Index was over 500 points below the previous Friday's close. There was nothing apparently remarkable about the intervening weekend.

16. In recent times, the term "hobbyist" might be expanded to include green enthusiasts. The early editions of the Toyota Prius were priced at a substantial premium relative to similarly appointed conventional cars. Prospects for recovering the premium via saved fuel cost over the life of the vehicle were not good, especially against the likelihood of a costly replacement of the battery pack at some point. Yet, greens were an important early market for hybrid drive autos.

17. An intense interest can induce personal sacrifices in favor of the center of interest. Witness the stories of Olympic athletes who worked for small incomes in order to enable themselves to adhere to rigorous training regimes.

18. The first modern hydraulic braking system is credited to one Malcolm Lougheed in 1918. The company bearing his name uses the revised spelling which he adopted: Lockheed.

19. Note the use of the word "necessary." Another condition is needed to generate a technology revolution, namely "sufficient." This other condition is the

existence of or prospect for a large-enough market to absorb the output of low-cost production once achieved.

20. ENIAC's development was government-financed, and its purpose was the calculation of range tables for tube artillery and aerial bombardment.

21. The availability of government as a large-volume customer was a large factor in nineteenth-century development of low-cost volume production methods. The pattern in which production was *indirect* in the sense that one produced specialized machinery for the large-scale production of technologically ordinary goods, such as small arms, encouraged the notion of *roundaboutness* in production that characterized the Austrian School economists, such as Spiethoff. In this view of capital, obsolescence of the product produced would render the capital plant employed in its production worthless.

22. ENIAC was designed for the purpose of calculating range tables for tube artillery and aerial bombardment. It had enough features that it is recognized as the pioneer of modern computing.

23. Such efforts are commonly termed "industrial policy."

24. How companies made use of this advantage varies. The options ranged from operating with fewer analysts to undertaking a greater analysis load with existing staff. One can assume the state of overall corporate prosperity influenced the nature of the response.

25. Interestingly, some of these "obsolete" technologies never fully disappear. As of this writing, there is a small group of audiophiles who are convinced that there is no better quality of sound reproduction than that produced by amplifier circuits based on vacuum tubes, LP recordings, and turntables. Most of the historic producers of this equipment have long since ceased its production, leaving, in some cases, small producers that are truly monopolies that price accordingly. A high-end turntable, for example, can be priced at a level that would buy a luxury automobile.

26. One of the casualties of Americans' embrace of the automobile was the demise of the electric street railway, a process that was completed in the 1950s. Now that auto traffic snarls have become commonplace, are many cities turning to dramatic new technology? It would not seem so, for traffic problems have led, among other things, to a limited revival of the electric street railway!

27. The notable exception to this rule today is the electric utility industry, which lacks the ability to store power intraday, and therefore must maintain enough capacity to accommodate peak demand. The industry has attempted to ease this problem by building capacity based on gas turbine, which can be switched on and off relatively quickly.

28. The replacement exercise in a normal technologically setting would be to replace a steam engine with another steam engine of improved design; indeed, steam engines had benefitted from incremental improvements since their first introduction in the 1830s. In the normal technology case, the replacement equipment could be employed with no change in the infrastructure. Replacement with diesel–electric locomotives required a much different infrastructure. Such a replacement verges on revolutionary, at least within the overall confines of railway transportation.

29. There were cases in which railroads maintained money-losing services because of the regulation-imposed difficulty of withdrawing from these businesses. This is discussed below.

30. A large railway system is organized operationally into a number of geographically well-defined divisions. Railroads were therefore able to confine obsolete technology to one or more operating divisions. This allowed them to concentrate obsolesced infrastructure to one or a few terminals, such as steam locomotive repair facilities. Such facilities serving other parts of the system could then be retired.

31. The WPB was disbanded following the war along with the regulatory structure that it enforced. Interestingly, the sacrifice of all technological considerations other than maintaining an uninterrupted flow of production is reminiscent of industrial management in the Soviet Union through its entire history, wherein the incentive structure faced by factory managers rewarded production increases and little else. One side-effect of these policies was that there was little reward and considerable risk to a manager who interrupted production to install more efficient plant even when such was available. This system naturally produced huge obsolescence by world standards.

32. One interesting example of Union thinking arose in the early 1960s when effective miniature two-way radios were introduced into the operation of rail yards. These supplanted a system of hand and lantern signals by which switchmen communicated with locomotive crews and tower operators and which were less than fully reliable. While the radios made the switching crews' jobs safer and easier, the relevant unions claimed that its members were now "radio operators" and demanded extra pay!

33. "Buys and Modernizes Steam Engines," *Modern Railroads* 4, no. 11 (November 1949): 81–86.

34. This recalls Thomas Watson's oft-cited observation that the world afforded only a tiny market for electronic computers. IBM did manage to overcome any effects of this miscall.

35. This contributed greatly to line workers' fatigue, and was at the heart of a serious labor turnover problem. Ford's early response was the famous five-dollar a day wage, which reduced the turnover problem and also gained Ford personally (and temporarily) the reputation of a humane industrialist.

36. The number of U.S. factory sales of automobiles rose at an average annual rate of over 30 percent between 1900 and 1923. Except for an apparent "spike" in 1929, factory sales were essentially level after 1923. Based on data published in U.S. Dept. of Commerce, Bureau of Economic Analysis, *Historical Statistics of the United States* (U.S. Government Printing Office, September 1975).

37. The hydraulic-controlled disk brake has become standard as of the early twenty-first century. European manufacturers adopted disk brakes in the 1920s for various reasons, well in advance of North American carmakers.

38. There were at least two circumstances behind this. First, a Federal agency, the War Production Board, had the power to disallow any technological development which it regarded as in any way novel; this was a measure

designed to minimize delays to production of known and proven designs, especially consumer goods. Second, production of many consumer goods was ceased for the war's duration, the most conspicuous being automobiles.

39. Indeed, the immediate aftermath of the war can be considered an industrial golden age for the U.S. People who decry "globalism" in the early twenty-first century often do not seem to be aware of the historically unique conditions under which this age occurred, and how these conditions have changed.

40. There is no better example than the Douglas C-47 transport plane, which became the DC-3, a workhorse of the commercial airlines that expanded shortly after the war.

41. The recession that started in 1973 was the most severe of the entire postwar era prior to that of 2007. In it, the unemployment rate peaked at approximately 9 percent compared with an average peak of approximately 7 percent for all postwar recessions, except for the 2007 downturn.

42. "Modern quantitative methods" rest heavily on the assumption of Gaussianity, and therefore probably underestimate risks growing out of high leverage.

43. The National Bureau of Economic Research (NBER) is a private economic research organization which in the past developed much of the design and applied theory that is incorporated in the National Income and Products Accounts maintained by the U.S. Government. The NBER has become a semiofficial dater of recessions. Generally, a recession is measured from a peak of economic activity to a trough of activity. A recession is a significant decline in economic activity spread across the economy, lasting more than a few months, normally visible in real GDP, real income, employment, industrial production, and wholesale–retail sales.

3

The Course of Normal Technological Change: Where We Are in 2010

When examining normal science...we shall want finally to describe that research as a strenuous and devoted attempt to force nature into the conceptual boxes supplied by professional education.
—Thomas S. Kuhn

The immediate impact of a technology revolution is to create new fields of incremental improvement across a wide spectrum of industries. This interrupts the tendency in a technologically unchanging or little-changing world to experience decreasing returns to improvement. A technology revolution can be thought of a reset of the process of improving a line or lines of technology. As of late 2010, the U.S. economy finds itself in a condition of normal technological development in which many if not most of the threads of technology development have grown out of the recent IT revolution.

There are several broad patterns in how a period of normal technology develops. First, the earliest improvements of a newly introduced technology often take the form of increases in power density. Here the term "power density" is used in a broad sense: it refers to increasing the power of the technology while simultaneously reducing the bulk of objects from which power is derived. In the early twentieth century, the term "power" literally meant horsepower derivable from devices such as IC engines and electric motors. In late twentieth-century terms, the term refers typically to the ability of integrated circuitry to deliver more capability out of ever-decreasing physical volume. Increases in power density account for much of the oft-noted tendency for the physical mass of the goods components of gross domestic product to diminish over time. Increases in power density commence almost with the initial introduction of a new technology but does not stop with the termination of the revolution.

A second pattern is the decline in the cost of applying new technologies (noted elsewhere). Indeed, a dramatic decline in application costs played a crucial role in triggering widening interest in the new technology, but declines in costs of application are ongoing after the revolutionary phase. Cost reduction in normal technological progress can be regarded as a widening and extending of the application cost decline that brought the technology out of its first-stage innovation process. For a stellar example, one need look no farther than digital imaging. This technology emerged commercially in the early 1980s, and it was during this decade that digital images became the standard for quality illustrations in magazines and other publications. In the mid-1980s, the equipment for producing quality digital images typically cost $40,000 and up, thus limiting the technology to commercial printing. By the end of the 1990s, however, digital image costs had fallen to the point where this technology has virtually taken over the consumer, commercial, and industrial imaging markets.

A third important pattern in normal technological change is that potential for improvement of any thread of technology appears to be ultimately limited. Not surprisingly, the more important improvements often appear early in the life of the technology, but not always. Incremental improvements need not be limited to the early stages of normal technological change following initial introduction, for normal technological progress across a wide spectrum of industries creates opportunities for improvements applicable in particular industries. However, there is a maturing process that affects any technology. The use of stationary reciprocating steam engines in industry generally died out, not because there was an absolute absence of improvement possibilities, but because alternative technologies having superior power density and lower costs had become readily available. When a technology has matured, it is vulnerable to replacement in a revolutionary episode.

Power Density

Increasing power density is shorthand for the tendency for technologies to become more useful even as they get physically smaller. The classic example of improving power density can be found in the evolution of the electric motor, which was the heart of the massive technological change that underpinned the prosperous years between

1900 and 1929. When motors first became useful, around 1885, they were bulky devices best suited for heavy industrial applications, such as traction motors for electric streetcars. Progress was in the form of reductions in the bulk coupled with increases in thermal efficiency. Several events convey the timing of this progress: the first complete electric street railway system was installed in Richmond, VA, in 1888. As power density improved, motors in increasingly smaller horsepower ratings became available. Units had become small enough by the early teens to be the basis for automobile starting systems, and by the 1920s, primitive major household appliances. Electric motors were the main enablers of the development of coordinated mass production lines in manufacturing that appeared in the teens and after.

The recent example of increasing power density is summarized succinctly in what has become known as Moore's law: that the number of transistors and related devices that can be placed on a semiconductor chip of fixed size doubles every eighteen months. The pioneer computers that can legitimately claim to the antecedents of the powerful and portable computers that have appeared since the 1970s were enormous by recent standards—ENIAC weighed over thirty tons. The revolutionary phase of IT growth (the phase that propagated a major surge in general economic growth) occurred in the 1990s for the most part, but development continues as there have appeared numerous small devices with processing powers that were probably not even dreamed of in the days of ENIAC. Normal technology continues apace at the time of this writing even though the economy is experiencing a serious posttechnology recession. As amazing as present IT is, it has nevertheless become commonplace.

Battery technology has been around in some form for over 200 years.[1] It has recently become a center of intense interest growing out of a perceived need for electric automobile development and an economical means of short-term power storage to be used in connection with intermittent sources of renewable electric power. The large obstacle to electric auto development is the low energy density of batteries as compared with gasoline: the best of contemporary electric car designs is high cost, heavy weight, and of limited driving range between recharges, all due to the low-energy density of batteries. To date, normal technological development has not accomplished much more than to provide batteries in limited automotive use, such as electric starting circuits.[2] Similarly, expansion of the use of renewable power sources such as wind and solar as base-load power sources depends on the development of economical means of intraday power storage. Flow batteries are examples of intraday storage technology that is presently available, but the capital cost of a large flow battery exceeds the cost of a completely new coal-burning power plant.

With the aid of government subsidies of various kinds, the research effort to increase battery energy density has become intense, but no dramatic breakthroughs appear imminent. IBM research, for example, has announced the goal of a seven-fold increase in the performance of the best of today's batteries—lithium-ion technology—possibly using lithium-air technology. Opinions as to when this goal can be achieved to the point of an automotive battery competitive with gasoline vary, but 2015 is one of the more optimistic of these. Today's battery research and development constitutes a test of the idea that dramatic improvement can be induced in a technology that has historically been resistant to it.

Cost

As noted elsewhere, a dramatic decline in the costs of applying technologies played a large role in triggering a technology revolution. However, subsequent learning curve cost reductions, including those due to large-scale in production, should not be confused with the breakthrough cost reduction(s) needed to trigger a technology revolution initially. Learning curve cost reductions are part of normal technological development. The learning curve begins when new technology becomes economical in some application and is available for entrepreneurially guided improvement.

The learning curve that follows a technology revolution is multifaceted. It includes cost reduction that results from increasing the scale of production of key elements of a given technology. Indeed, falling costs growing out of a learning curve is often a familiar concept such that there are a number of instances wherein it has been incorporated into pricing decisions *in advance* of the realization of cost savings from large-scale production. There was a famous instance of this pricing logic in the early history of the transistor, in which Texas Instruments priced transistor sets for radios at below cost in a decision deliberately aimed at creating a market for these radios *before* mass production facilities were on line. The expectation was that introduction of large-scale production would bring costs below price. Actually, pricing strategies of this kind have not been unusual in the so-called high-tech industries.

The Model T experience is far from the only example of falling costs due to increasing production scale. From the earliest years of the automobile industry, substantial improvements such as electric starting and hydraulic brake controls have been introduced at the luxury end of the price spectrum. In two early examples, electric starting was first offered by Cadillac in the teens, and hydraulic brake controls first appeared on the 1921 Duesenberg. These and other improvements began to be offered downscale as design and production efficiencies improved. To a limited extent, features that at one time were marks of luxury were "democratized."[3]

To appreciate the size of the cost problem, it is useful to return to the electric car possibilities. The Nissan Leaf, scheduled to be in dealer showrooms in the fall of 2010, will have a sticker price of $33,000. This model is basically a $16,500 model with a battery pack that costs $16,500. As things presently stand, it is difficult to envision this model as a "peoples' car"; but it could appeal to wealthy consumers who are emotionally attached to contributing to air cleanliness. Sales to such a group would be reminiscent of the market for automobiles before 1905.

There are a number of present-day examples that illustrate the power of scale. The array of hand-held electronic devices that offer capabilities ranging from photography, game playing, and Internet access is little short of remarkable by standards of only a few years in the past. Amazing as these devices are, however, they represent evolutionary, not revolutionary, technology development. They are clear lineal descendants of the prototype inventions such as integrated circuits. Their stage-one innovation was largely financed by government, and in many cases in the context of defense-related projects. The scale that creates the economic incentive for improving these devices rests on popularity: sales have been in the millions, sufficient to justify development of whatever specialized production technology that has been necessary to move to each step of technological improvement.

Technological Maturity: Rising Costs of Incremental Improvements

Many of the most important improvements that normal technology will provide occur fairly soon after the close of the revolutionary phase. There are clear exceptions to this "rule," for even normal technological progress will produce enabling technologies in one sphere that foment at least brief accelerations in another line of technology. However, there comes a point at which a technology is mature in the sense that large improvements have become physically very difficult or costly to achieve. Historically, one has only to look at what happened to the stationary steam engine after 1900. In the fifteen years prior to that year, a number of more efficient power technologies had appeared, and producers of steam engines found themselves increasingly unable to compete with these on the bases of the incremental improvements that they were achieving. But for a few *niche* markets, stationary steam engines were totally obsolete by 1930.

The effective definition of technology maturity is a condition wherein "improvements" come to have a cost in excess of total benefits flowing from them. If an alternative is available, then the mature technology is vulnerable to replacement. This is illustrated in Figure 3.1. The more time that passes, the economic impact of the next improvement of a given technology diminishes while the total cost of achieving it increases. In Figure 3.1, improvement effort achieves a maximum at point "A," at which costs of incremental improvements begin to exceed incremental benefits.[4] If this process continues indefinitely, the result is technology stagnation. The role of the technology revolution can be

Figure 3.1
Maturation for a Specific Technology

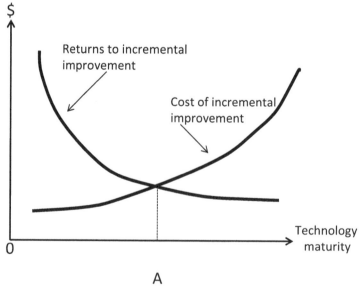

understood as a resetting of the normal technology process, providing it with a renewed slate of improvement possibilities.

The point "A" at which returns to incremental improvement can be thought of as optimal in a cost sense, but may not actually be the limit to which a technology is actually pushed. The physical potential of a line of technology may be pushed well beyond the optimum if it is subject to political influence. A current case in point is the movement to eliminate greenhouse gas emissions, especially carbon dioxide (CO_2). There are a number of issues here, not the least of which is the fact that the magnitude of costs associated with rising levels of atmospheric CO_2 is not easily measurable, and therefore is controversial, and as a result, there is no scientific or popular consensus regarding where the optimum point is.

When a technology is mature, it is vulnerable to replacement by a better technology *should one be available*. The fate of silver halide imaging (film photography) is a recent case in point. Up to about 1990, silver halide imaging was undergoing improvements, but none of the improvements had more than minor impact of the costs of or benefits for users of the technology. The rapid improvement of digital imaging offered the possibilities of substantial cost reductions in cost with little or no quality sacrifice of image quality. Amateur photographers are well aware of the abandonment of film photography, but the major impact was in the health care industry. For years the big-volume market for silver halide imaging was X-ray plates. These required processing via a chemical process and results of exposures were available for diagnoses only after a delay for processing. Digital images can be created and the results viewed almost instantaneously, and the processing facilities needed with the silver halide technology could be dispensed with. This change has affected almost every hospital, medical clinic, and dental office.

The almost total capture of the imaging market by digital processes was made possible by the invention of the CCD. This is a device for the movement of photonic electrical charge, usually from within the device to an area where the charge can be manipulated. The CCD originated from Bell Labs in 1969. Inasmuch as digital imaging based on the CCD had virtually captured the market for high-end commercial illustration by 1985, the suggestion is of a rapid stage-one innovation process. Subsequently, falling costs relative to quality imaging enabled CCD technology to take over almost all imaging markets by 2000. The CCD's technology antecedents include miniature integrated circuitry and photoelectric technology.

Where are we Now?

There is no sign of maturity in the IT field as of late 2010, but there is also little suggestion of any development likely to trigger an investment boom sufficient to support the kind of rapid economic growth of the 1990s. Where are the applications of IT appearing? It is easy to find examples of ingenious applications that take advantage of the continuing advance according to Moore's law. Microprocessor circuits continue to enable even more activities to be encapsulated in hand-held devices. The present trend is for hand-held devices to include Internet access. Moreover, these devices are increasingly gaining the capability for downloading very large files at speeds that enable downloads to take place in short time intervals.

Yet most of what are commonly called high-technology developments as of 2010 can be characterized as normal technological progress. By this is meant that there is little presently happening that has the potential for really large changes in the way goods and services are produced and how consumers live. The new devices that appear in consumer and other markets sell well enough to support a handful of large enterprises that produce and sell them, but very little of this technology can be said to auger the kind of changes comparable with what occurred during the revolutionary phase of digital technology, as during the 1990s. What is happening today is fully consistent with the normal technology characteristic in which private sector inventive and innovative effort is concentrated on improving existing lines of product, not on stage-one development of something radically new.

Notes

1. The term "battery" comes from Benjamin Franklin, who ranks as a pioneer in the electricity field. He seems to have inadvertently grasped the two leads of a charged leyden jar, and likened the effect to a severe beating, or the legal term, "battery."
2. It would be interesting to see if a technology revolution, which a battery of dramatically enhanced power density would amount to, can be induced by throwing money and talent at it.
3. If luxurious features eventually offered through the automotive price spectrum, then why does the luxury car market still exist? There are a number of reasons, which interestingly include some engineering ones. One example is silence. The suppression of road noise has never been achieved except at

the high end of the price range. This technology is costly, and it has never been clear to carmakers that the market for less-costly models would support such a change.

4.   Readers who are familiar with intermediate price theory will recognize Figure 3.1 as an adaptation of a chart used to illustrate the relation between the value and costs of pollution control. The idea presented here is much the same, but the rising costs and diminishing benefits here reflect diminishing returns in the maturation of a technological paradigm.

4

On Recognizing the Technology Cycle

Facts do not cease to exist because they are ignored.
—Aldous Huxley

Introduction

In the late years of the first decade of the twenty-first century, no one would argue that the march of technology has in any way ceased. Within very recent memory, the plain cellular telephone (good for making voice calls) has morphed into devices that not only can take pictures, but can access the Internet. Is today's technology any different from that of the late 1980s and 1990s? From the standpoint of economic growth, the answer is emphatically yes. Recent technological miracles represent no more than incremental developments of the hardware and software of the IT revolution: normal technological change. By contrast, the hardware and software that grew into today's marvels were novel in 1990. Their general adoption by businesses and consumers was waxing fully. Expanding demand was inducing the construction of new facilities for producing the IT hardware and software. In 1990, the tools of IT were revolutionary. In 2010, they were commonplace.

Inasmuch as the state of the technology cycle very much influences the nature of the effects of conventional monetary policy tools, it is highly desirable that the tech cycle be closely monitored. Unfortunately, the statistical means for doing this are far from satisfactory. There are several probable reasons for this. Of these, the most important is that ever since the great depression of the 1930s (and before), serious recessions (and worse) have been regarded as due to unstable financial market conditions; little attention has been given to possible causative elements in the real side of the economy. Consequently, research and development of the general statistical bases for monitoring the condition of the economy has neglected real events. Collapses of capital investment flows have been regarded as consequences of financial market failures; hence the idea that these flows can be restored by means of financial market fixes, aided by fiscal spending in the worst cases. The possibility that investment in productive assets can fail due to a temporary paucity of productive investment opportunities generally is not considered in present-day government policy decisions.

Still, policymakers are not totally bereft of evidences regarding the technology cycle. The emergence of a prototype invention, discovery, or an equivalent event such as a radical change in a regulatory regime, can probably be regarded as an unpredictable event. The same is especially true for a combination of these events in a short span of years. However, once the underlying principle of the prototype(s) has entered the pool of knowledge, the general direction(s) of inventive activity that moves the prototype idea into a commercially useful form is predictable, at least conceptually. Therefore, it makes sense to investigate the question of how the progress of the idea through its innovation stages can be followed in a systematic manner. Resulting understanding of the innovation process is necessary if there is to be any useful policy response to the wider economic development events that stem from the innovation process, especially in the second stage of innovation.

How such clues might emerge and be read shows in the example of the IT revolution from whose stage-two innovation came the investment-led prosperity of the 1980s and 1990s. The elements underpinning the IT revolution go back at least to the invention of the vacuum tube shortly after the turn of the twentieth century, and, in some cases, even further back to such inventions as punch card input around 1800,[1] and the concept of the modern computer, credited to Charles Babbage in the 1830s. The machines that are recognizable as modern computers date from the 1940s. The earliest of these (e.g., ENIAC) tended to be built as special-purpose machines, but the early computers evolved into generalized programmable machines that offered increased efficiencies to standard business activities, such as payroll accounting, and thereby justified continued development. The stage-two innovation of IT can be dated from the early 1980s, when falling costs and expanding capabilities of computing made it economically attractive, to apply computerization to an expanding array of increasingly complex business problems.

Several conditions are common signatures of technology revolutions. These include falling costs associated with applying the new technology, the formation of new enterprises in new industries and the growth of some of these enterprises to large size in a comparatively short period of time, and a surge in the rate of general productivity increase. Consider the emergence of the mass-produced automobile in the early twentieth century, which was the basis for dramatic reductions in the cost of travel and transport over land. A similar reduction in transport costs resulted from the nineteenth-century development of railroads, and the "age of steel" which had its heyday in the late nineteenth century and grew out of technology-based falling costs of that material. The obvious recent example is the dramatic drop in the cost of electronic computing in the 1980s and 1990s and the consequent explosion in the applications thereof.

When the new technology is based on a cluster of prototype inventions that appeared in a comparatively short historical time interval, the range of possible applications is wider than the simple application of any one of the new inventions/discoveries to doing more efficiently what had previously been done. Also, the range of possibilities that the appearance of a slate of prototype inventions presents is not always immediately recognizable because the outcomes will stem from combinations among the new and preexisting technologies. Realization of the entire range of possibilities is subject to a learning curve which can take decades to evolve. However, when the wider implications of a technology revolution have begun to appear, it becomes possible to incorporate the fact of it into policy decisions.

Signatures of the Upswing

Falling Costs

Technology revolutions are not born in a vacuum; they are typically outgrowths of technologies in stage-one innovation. The event that most commonly triggers the revolution is some engineering breakthrough that brings a dramatic decline in the costs of applying the technology(ies). When such a cost decline can be identified, the fact of it constitutes a leading indicator of an oncoming revolution. The most striking thing about the comparison of Figure 4.1 is the contrast between the downward movement of semiconductor prices and the

Figure 4.1
Producer Price Indexes

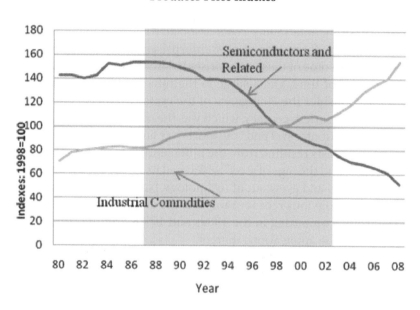

Source: Bureau of Labor Statistics. Shaded area corresponds to technology boom.

increase in the industrial commodities price index. The semiconductor price decline is consistent with an overall decline in the costs of applying computer technology to a wide spectrum of industrial and other economic activity. It was this widening application of computer technology that was the object of the investment boom that underlay the general prosperity of the period. The semiconductor price decline is consistent with an overall decline in the costs of applying computer technology to a wide spectrum of industrial and other economic activity. It was this widening application of computer technology that resulted in the investment boom that underlay the general prosperity of the period, especially the 1990s. The shading denotes the approximate duration of the IT revolution.

The reader will note that the semiconductor series did not begin to fall until the early 1990s. This suggests that a sharp decline in the costs of applying a technology is not a leading indicator, for the 1990s was the decade of the investment boom. However, it would have been an effective *coincident* indicator, and this would have made it useful for providing insight regarding how quickly recovery following a financial market setback can occur. Recovery could be quick if there remained unexploited investment opportunities in an ongoing technology upswing.

There is no better and succinct characterization of the IT revolution than Moore's Law. This states that the number of transistors (and other devices) that can be placed on a semiconductor chip of given size doubles every eighteen months to two years. As has often been remarked, this is not a "law" of nature in the sense of the law of gravity, but is an indicator of the rate of advance of a powerful technology, and the possibility of its continuation into the future is not guaranteed except by the belief that technological progress can continue to deliver past rates of growth. What is not remarked as often as Moore's law itself is its implication for the costs of computing and other applications of digital technology. The massive expansion of digital technology rests on the falling costs of its application. As in the first decades of the twentieth century, the technology revolution grew out of a mass production process: that of digital-based circuitry and high-grade silicon.

Could this kind of evidence have been discerned in earlier instances of technology-induced investment booms? A direct reproduction of the comparison of Figure 4.1 for the pre-1930 years is not possible, for while the producer price index (PPI) for industrial commodities is available for years 1913 and onward, there is no directly comparable series bearing directly on the new-technology growth industries of the time, automobiles, and electrification of factories and homes. An understanding of the falling-cost evidence of that time depends on other evidence. Figure 4.2 shows part of the picture, namely the history of prices of industrial commodities. While not specific to pricing and cost events in the new industries of the times, this index will reflect the *overall* price effects of declining production costs of a variety of growing major industries. The shaded area denotes the approximate years of the technology revolution that revolved around the automobile and electrification of the economy. The year 1923 was chosen as the late limit because that year was the break point between very rapid growth and no growth of automobile factory sales.

The rising price index prior to 1920 reflects the period in which mass production methods were being introduced. While these production methods eventually came to dominate the overall index of industrial commodity prices, it may be supposed that they did not

Figure 4.2
Wholesale Price Index: Industrial Commodities, 1913–1930

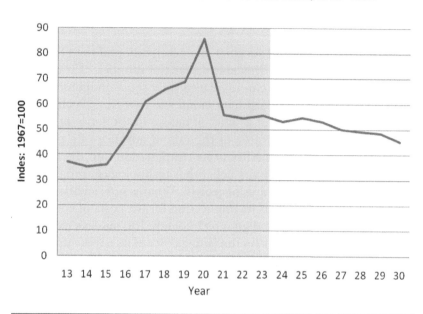

Source: Historical Statistics of the United States.

dominate the overall index until well after 1913. Moreover, there were other influences tending to push up industrial goods prices during this period, such as a surge in demand from government related to World War I. In 1912, the list price for a Model T Ford was $875.00. In the year 1926, just before the end of Model T production, the list price was $270.00. This by itself does not justify the assumption that the Model T pricing reflected the entire auto industry of the time, but two other facts do uphold such an assumption. First, Ford's production methods were coming into more general use by 1920. Second, the Model T enjoyed 60 percent of the new car market in 1921, which strongly suggests that had a PPI for automobiles been published during that period, Ford's pricing policy would have weighed heavily in it.

There are important advantages to being able to cite government statistical evidence as applied to a question, such as which industries are distinguished by falling costs. These include the trust that is commonly accorded to these data. As this discussion suggests, however, there are problems that grow from relying on government for unmistakable evidence by which falling costs situations can be recognized. Perhaps the biggest of these problems is that in order for a government statistical agency to justify the cost of producing a price series that is relevant to an emerging strain of technology, it is necessary that the agency recognize the public importance of the information. Once this happens, design of the data can be a time-consuming process. The result is likely to be the best that statistical professionalism can produce, but it may not be available for general use until the trends depicted are fully established; or worse yet, until the rapid early growth period of the new technology is past. The price indices for semiconductor products depicted in Figure 4.1 go back to 1980, but another series which would be of interest, the PPI for computers and related goes back only to 1990. Clearly, if one wishes to study the technology cycle over a long reach of history, other evidence is crucial.

New Industries

One salient characteristic of technology revolutions is the emergence of new combinations for doing things not previously done, and these are commonly domiciled in new industries.[2] Therefore the emergence of new industries and the attainment of large size of some of the firms within the new industries is a strong indicator of a technology revolution. The classic example is the array of industrial and infrastructure development surrounding the automobile industry in 1900–1920, and the recent example is the massive expansion that falls under the heading of IT. In recent times, think of Microsoft, Intel, Oracle, Cisco, and others, new firms which have grown to large size on the basis of the basic inventions underpinning the IT boom: the transistor, the integrated circuit, fiber optic cable, and the laser.

The emergence of new industries is initially marked by formation of new companies, of which a few grow rapidly to large size. This period of initial industry expansion is often accompanied by consolidation, including mergers and acquisitions leading to a reduction in the number of firms in the new industries. The reasons for the consolidation vary from cycle to cycle, but generally speaking, the consolidation phase accompanies an initial period of spreading of applications of the new technology that was the basis for the expansion. In the early twentieth century, many consolidations were motivated by problems arising from the sometime major size disparities between large firms and firms that supplied them. The result of this consolidation was vertical integration. Salient examples include Ford Motor and Sears Roebuck. In the late twentieth century, the consolidation phase of the expansion was motivated by desire of the technology companies that had attained substantial size to acquire technologies possessed by smaller companies.

The appearance of completely new industries is the most disruptive feature of technology revolutions as compared with normal technology. As noted above, a failure to find a single technological change that dramatically changes the nature of a single industry's progress *does not* justify a conclusion that there are no technology revolutions, for such conclusions in the past have been based on studies of *established* industries and have nothing to say about new industries. Technology revolutions are *inter-industry* phenomena, not intraindustry.

The Productivity Signature

A third signature of the presence of a strong technology push to a general economic upswing is an above-normal rate of general productivity growth. This is a result of the investment surge in which older capital plant gets replaced by new plant having superior productivity characteristics. Figure 4.3 shows a history of productivity since 1915. The raw ratios have been smoothed via a moving average procedure in order to dampen the year-on-year tendency for productivity ratios to be excessively volatile, especially in the first two decades of the twentieth century. Volatility in the year-on-year productivity comparisons can occur in response to short-term financial market crises and other events such that the smoothed productivity series better reflects the long-term effects of the underlying investment boom on growth.

Inasmuch as there is no generally accepted index of technology change, it is not surprising that productivity change has been a focus of many studies of its impact. As a proxy for technological progress, however, productivity growth has some problems. Among these is it measures change in a *ratio* of output to factor input. The idea is that technical progress results in increases in the numerator relative to the denominator. However, this occurrence does not always reflect what one is likely to consider to be progress. How does one interpret the prolonged increase in productivity growth that commenced in the mid-1930s and did not fade until the late 1960s and into the 1970s (see Figure 4.3)?

Figure 4.3
U.S. Real GDP per Worker-Hour: Ten-year Centered Moving Average

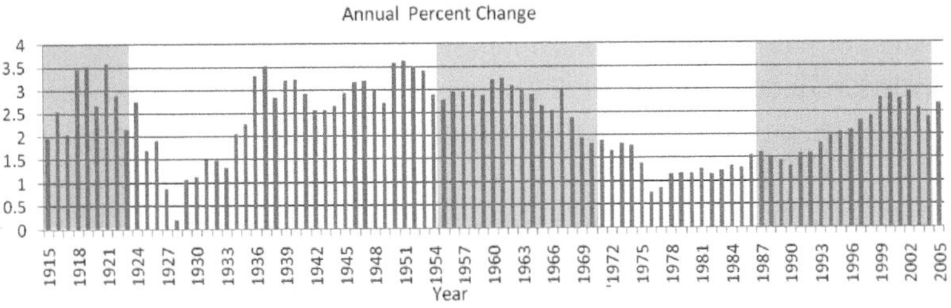

Annual Percent Change

Source: U.S. Department of Commerce, Historical Statistics of the United States and Bureau of Labor Statistics Shaded areas represent approximate investment booms.

The smoothed productivity data eliminate some, but not all the volatility in productivity ratios that stem from short-term economic events taking place atop the underlying technology cycle. The depression-era increase in productivity growth (especially after 1935) does not reflect expanding output (numerator of the productivity ratio); it reflects the reduction in production cost elements, especially labor input (denominator). The depression lasted long enough that all costs became variable, and could be reduced in proportion with diminished revenues. In order to cope with depressed demand that characterized the depression, firms sought to reduce their costs to a level commensurate with reduced sales; and this cost reduction effort was well along by 1935, when productivity growth appears to have accelerated. Generally speaking, a productivity growth surge can *verify* the presence of a technology basis for a general growth surge, but the reverse logic is not valid: one cannot infer a technology push to growth from a productivity growth surge in the absence of corroborating evidence. The growth in productivity in the depression period patently cannot be interpreted in this manner.

High levels of productivity growth during the war years were achieved under pressure of rapidly expanding wartime demand and worker shortages. Continued high levels of productivity growth in the 1950s and 1960s were aided by a stage-two investment boom based on technologies whose origins in some cases dated into the depression years and which were developed in the war years. These technologies became widely available to the civilian economy in the decades after the war and underpinned the general prosperity that characterized the decades up to 1970. Examples include commercial jet air travel, consumer electronics (especially television), and nuclear power.

There are clear surges in the productivity ratios in the years prior to 1923, to 1970, and since 1990. These surges correspond to the impacts of, respectively, the investment booms centered on the automobile and electric power industries, the postwar prosperity, and on the IT boom. These intervals are indicated by shadings. It is of interest to take a close look at the recent productivity change in relation to a statistical record of its principal driver: investment in information processing technology and software. Annual percent change in this investment is shown in Figure 4.4 juxtaposed with the change in smoothed productivity. The shaded area denotes the approximate IO revolution.

The striking feature of this comparison is the surge in investment that took place between 1991 and 2000, and the contemporaneous increase in the productivity growth rate. In this decade, productivity growth increases. This chart ends with 2003 because of the smoothing of the productivity growth data; and it is reasonable to ask what has happened since. From 2004 to 2008, productivity growth averaged 2.34

Figure 4.4
Real Investment in Information Processing Technology and
Software Growth in Overall Productivity (smoothed)

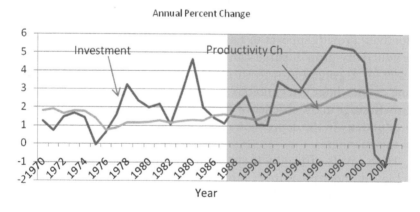

Source: U.S. Bureau of Economic Research and Bureau of Labor Statistics. Shaded area denotes technology-driven investment boom.

percent per year and was 1.7 percent in 2008. Investment growth in the same period averaged 1.85 percent per year and was 1.14 percent in 2008. These numbers are consistent with continuation of a decline in the rate of productivity growth following the waning of a major technology-founded upswing.

The decline in the productivity growth rate after 1999 reflects a complex process of which an important element is the maturing of the IT-driven upswing. In the mid-1990s, IT-related investment replaced pre-IT technology. In the early 2000s, it was replacing earlier IT-related technology, and, not surprisingly, the impact on productivity was less. The decline in the smoothed productivity ratio's growth is a clear sign that the growth impact of IT-related investment was waning. Inspection of Figure 4.3 reveals that there was a similar decline in the rate of productivity increase in the mid- and late 1920s, and in the late 1960s. In both of these instances, a pronounced decline in the smoothed productivity growth ratio heralded a general economic collapse of some sort. The recession of 2000 diverted attention from the general weakening of the investment surge that drove growth in productivity in the 1990s.

On the Appropriateness of Output per Worker–hour in the Study of Growth

Past study of productivity has been closely associated with the concept of the aggregate production function, a simple mathematical expression which purports to relate inputs of factors of production (usually labor and capital) to output for an entire economy. Applied to a narrowly defined production process, the production function proved useful in comparisons between alternative technologies, as each production function corresponds to a particular technology. Interest in production functions applied to entire economies (aggregate production functions) grew from interest in explaining the distribution of income as between owners of capital and providers of labor services. However, aggregate production functions have not proved especially useful as tools for the empirical study of overall economic growth. The main reason is the capital measure used: these measures as typically constructed do not reflect changes in the productive *quality* of capital. The actual capital stock is always an evolving mixture of technologies of differing productivity characteristics. For various reasons, no measure of aggregate capital stock has to date succeeded in capturing the concept of evolving quality change in aggregate capital.

This study has relied on productivity defined as output per unit of labor input for three reasons. The first is that data reflecting this concept are available in a much longer history than alternative measures, such as total factor productivity. A second reason is measures of total factor productivity are based on hopelessly flawed measures of capital stock. The third, and most important, reason is that output per worker hour does an adequate job of capturing the relation between output and technological progress, but not always contemporaneously. As measured, the productivity ratio appears to reflect contemporaneous relations between X and L; however, growth in X will reflect changes in technology of the past as well as contemporaneous ones. It is not generally possible to distinguish between productivity changes due to contemporary events and those of the past. Consequently, the X/L ratio will pick up lingering past productivity changes sooner or later.

In an example, after 1995, railway investment concentrated heavily on elimination of bottlenecks to rapid train movement in the form of line straightening, second main tracks, and other physical plant improvements. The object of these improvements was to upgrade the railways' ability to deliver reliable and rapid service and was primarily a reaction to highway freight competition; the immediate impact on employment was minimal, as was the impact on labor productivity statistics. Yet there was an impact on employment. By increasing the rate of capital turnover (less rolling stock needed to maintain service because of more rapid capital turnover), fewer workers in activities such as car and track repair would be needed eventually.[3] A reduction in the output-to-labor force will affect some year's statistic regardless of whether or not the cause of the reduction occurred in that year.

Technology Revolution: The Energy Signature

There are two signatures of this progress. The first is overall productivity in the economy. The other is the signature that technology revolutions leave on the energy efficiency of the economy. The energy efficiency of the U.S. economy, expressed as energy consumed per dollar of real gross domestic product, has consistently declined since about 1920, as shown in Figure 4.5. Inspection of Figure 4.5 reveals that the rate of increase in energy efficiency has not been perfectly uniform through time: there were distinct periods in which the decline in energy consumption appears to be more rapid than in other times.

Figure 4.5
U.S. Energy Consumption per Dollar of Real GDP

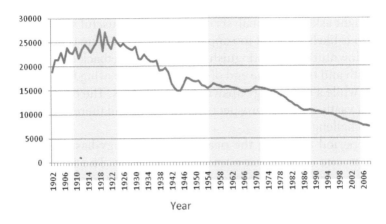

Billion Btu per 2005 Dollar

Year

Source: Energy Information Agency, Bureau of Economic Analysis, *Historical Statistics of the United States* (1973), Schurr *et al.*, *Energy in the American Economy*, 1960.

The shaded intervals in the figure mark the episodes in which the improvement rate in energy efficiency appears to be less rapid, or even declining, as before 1922. Interestingly, the shaded intervals are approximately those that are identified in previous chapters as periods of technology revolution!

Why should a technology revolution be temporally associated with either an absolute decline in overall energy efficiency (as in the pre-1922 period) or a diminution in the rate of energy efficiency improvement? One hypothesis is that in a technology revolution, transforming new technology is relatively crude in comparison with its eventual improved form. The main activity in the revolutionary phase is the *spread* of the new technology across the economy. Spreading the novel technology across a wide swath of the economy draws resources of talent and financing from the normal technology activity of improving incumbent technology. Figure 4.5 suggests that the technology revolution is a discontinuity in the course of normal technological change, for it radically changes the array of technology threads on which normal technological change works.

Signatures of the Downturn

Downturns following two notable technology upswings, those following 1929 and 2007, were abrupt. The downturn that followed the prosperity of the 1960s was not conspicuously abrupt by comparison. One might ask why an analysis of downturns is useful when the fact of them is likely to be all too apparent. The main issue is not the fact of the downturn, but distinguishing a financial setback that is associated with and triggers the end of a technology upswing from one that occurs while a technology upswing is in progress. The two situations present vastly different prospects for recovery and therefore different policy challenges.

The period following the peak of a technology-based economic boom is one of high economic vulnerability. The basic reason for the upswing was the expansion of demand for the products of the new technologies, but the rate of expansion diminished with the maturation of the markets for the new products and the new productive systems which they enable. A financial crisis can trigger the collapse, and the policy reaction to the crisis can affect the ability of the economy to recover adversely if it does not specifically recognize changes in the economy, especially labor markets, wrought by the preceding technology-based investment boom.[4] The financial system can be thought of as resembling a superstructure resting of a real foundation whose condition is sensitive to the rate of technology flux. The full exploitation of a slate of investment opportunities that grew out of an especially fecund set of prototype inventions and discoveries is equivalent to serious change in the nature of the foundation.

There are five specific conditions that point to a downturn's coinciding with a downturn in the technology cycle that grow either out of a general seeking of high yield, or temporary disappearance of opportunities for productive investment: a sharp decline in net new business formations; general inflation; change in the overall composition of demand for goods and services; asset price bubbles; and fraud. The fourth and fifth of these grow out of an abundance of liquidity and the pursuit of high returns in the absence of productive investment opportunities. Productivity growth data also offer a clue to diminution of a technology revolution: a reference to Figure 4.3 above reveals sharp declines in the productivity growth rate in years following 1923, 1966, and 2000. These declines reflect the transition from revolutionary technology to normal technology, as the products of the technology revolution become commonplace.

Sometimes, one or more of these five conditions will appear in the form of a recession, one that occurs while there are still productive investment opportunities from the boom on which economic prosperity had been feeding, but which nevertheless gives warnings regarding coming exhaustion of productive investment opportunities. The recession of 2001 is a classic example. It involved a failure in an important sector of the IT revolution, namely telecommunications infrastructure. This was a clear sign of the maturing of an important market for technology products, coupled with overinvestment in the sector, even though other components of the high-technology industries were less affected. While detailed comparisons are not easy in the 1920s because the structure of the economy was very different from what it is at the present, a series of relatively short recessions related to weaknesses in agricultural product prices contained hints of the problems that crushed the economy after 1929.

Fading Productive Investment Opportunities

It was pointed out above that a technology revolution is characterized by the formation of new companies which typically play a large role in exploiting the business possibilities of the new technology. The transition of technology revolution to normal technology is reflected by a diminution of the slate of productive investment opportunities that fueled the formation of new businesses during the revolution, with a consequent diminution in net new business formations. Inasmuch as official data on new business starts and business failures have been published only since 1992, a comparison of starts and failures before that year can be accomplished only indirectly at best. However, the data since 1992 do provide a clue regarding the behavior of net business starts since 1992. Figure 4.6 shows a sharp decline in net new starts following a peak in the first quarter of 2006. The shading denotes the approximate period of the IO revolution. Serious difficulties in financing businesses, especially small businesses, developed along with the financial crisis that developed in 2007, and there has been a tendency to regard the decline in net starts as due to adverse financial market conditions. However, the decline in net starts commenced in 2006, *before* the general financing problems became evident. This is consistent with the idea of a growing paucity of productive investment opportunities associable with the end of the technology boom.

Figure 4.6
U.S. New Business Starts and Failures

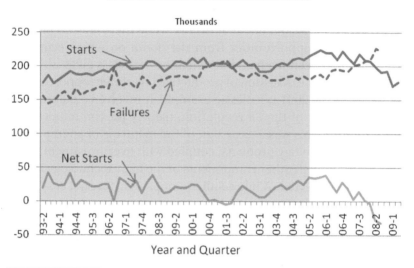

Source: Bureau of Labor Statistics, Business Employment Dynamics. Shaded area is approx. tech. revolution.

The suggestion is that a sharp decline in net new business starts is an indicator of the fading of the slate of investment opportunities that characterize the technology revolution.

Productivity

Productivity change is a useful indicator of the waning of a technology revolution and warns of the weakening of the investment boom. The fading of a technology revolution evidences itself in declining gains in overall productivity. This reflects that as adoption of the revolutionary technology spreads in the overall economy, the "new" technology evolves into normal technology. This can be seen in the historical record, in Figure 4.7, which presents productivity change in the form of unsmoothed annual data. Immediately following, each shaded interval, the rate of productivity change falls substantially and remains depressed for several years.

Willem Van Zandweghe, in a recent study, has noted that the cyclic performance of productivity in relation to output has undergone an apparent change since about 1984. Prior to the mid-1980s, output and productivity were approximately correlated over the course of the

Figure 4.7
U.S. GDP per Worker-Hour: Year/Year Percent

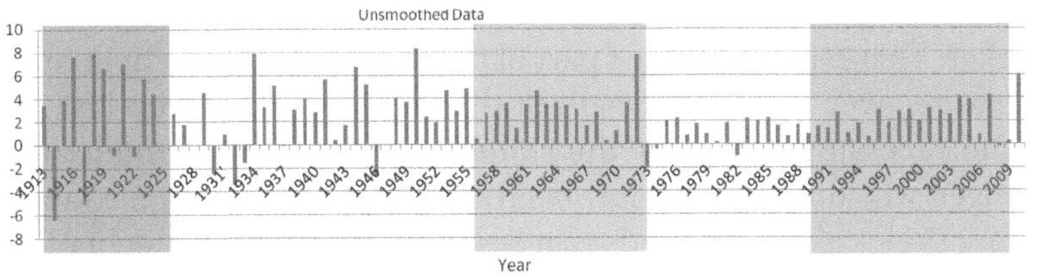

Source: Bureau of Labor Statistics, Historical Statistics of the United States.
Note: Shaded intervals are appoximate dates of technology revolutions.

business cycle, but since then, this level of correlation has considerably diminished. In the post-2007 recession, output and productivity took sharply different directions: with productivity sharply up and output sharply down. The author concluded that structural changes in labor markets were a large part of the explanation.[5]

This finding is consistent with the jobless character of the recovery, discussed immediately below; however, it is also approximately consistent with another trait of the aftermath of a technology-based investment boom, namely that while the rate of productivity growth declines as the boom wanes, it tends to *increase* in the recession following the boom. This shows better with the smoothed productivity data (Figure 4.3) than it does in the unsmoothed data (Figure 4.7). What appears to be happening is that during the boom, the productivity ratio is dominated by its numerator, but as the recession gains dominance in companies' thinking, the influence shifts to the denominator. When firms grasp that they are facing a serious recession wherein prospects for short-term business expansion are dim, they turn to cost-cutting measures, usually including layoffs. This results in the large jumps in productivity that Figure 4.7 shows in years such as 1934–1936 and 2010. The weak "recovery" that has taken place since 2008–2009 has been notably without much effect on the unemployment rate, but it has produced some large single-year jumps in the productivity ratio. A growing deviation between the directions of productivity growth and that of output growth appears to be at least a coincident indicator of an oncoming deep recession, and may even be somewhat leading.

Unemployment

A technology-based growth impulse brings changes in the organization of business one of whose consequences is a changed labor market. This explains at least part of the failure of the unemployment rate to recover in the aftermath of the boom: workers who lost jobs in the recession may lack the skills demanded by an industrially changed world, or have simply been surplussed by companies' ability to function with fewer hands that resulted from the new technology. If a recession occurs during an ongoing technology revolution, then employment recovery will probably take place relatively quickly in response to monetary policy actions. Such actions are effective because the unemployment is mostly a cyclical phenomenon. If, however, the recession follows the end of a technology-founded investment boom, or occurs in its waning years, monetary actions that are designed as purely countercyclical measures will not be effective, as this unemployment is *structural*, not cyclical.

Recovery in the employment rate has been the subject of much study, and the "jobless recovery" has attracted much scholarly attention in the past two decades. There are two bodies of explanation for this phenomenon: (1) globalization and (2) technological change. Of these two, the latter has probably received less attention than it deserves, for while some impact of technological change has been acknowledged, there has been almost no discussion in the literature of variations in the *rate* of technological change that might affect the rate of employment growth.

Perhaps the most obvious impact of globalization is on manufacturing activity, whose share of U.S. GDP has been declining for a number of decades. It is fair to ask what is the evidence that a surge in technology is also a contributor to the unemployment problem? The pattern of growth in productivity in the 1980s and 1990s may offer a hint, for in this period, productivity expansion went well beyond goods-producing industries; it affected the service sector profoundly. The IT revolution affected how services providers are organized. Changes in how these industries operate lie at the heart of their ability to operate with fewer employees, including substantially reduced numbers of middle-management and professional workers. Many service industries—for example, those involved in transportation and distribution—are not subject to offshoring to the extent that goods industries are. In a service business, contracting for output from a foreign factory is a much different and more difficult problem than contracting with a foreign factory to produce goods. Therefore, offshoring is an unlikely explanation for lingering unemployment of service-area workers.

Jobless recoveries are nothing new. The recovery from the 2001 recession was of this character. Corporate profits and investment did recover, but for the overall economy, there was only a fading of the technology impetus that had grown from the IT revolution. In this fading, business growth prospects were disappearing, and absent these, there was no strong incentive to take on new employees. Technological progress was reverting to the normal, and in this condition, investment reverts to the level of replacing and modernizing existing capacity. Other components of investment needed to support a growth impulse are temporarily gone. These include the need to provide infrastructure to support new technology, replacement of obsolesced productive capacity, and others. If the long reach of history is considered, it can be argued that the ultimate jobless recovery was that between 1933 and 1938.

In both recent and historical cases, the problem that a paucity of growth-inducing private investment opportunities poses has not received even remotely as much attention as it deserves. What is coming into view is the uncertainty in the business climate created by the government reaction to the recession. This compounded with a poverty of growth-inducing investment opportunities is a deadly combination for private investment. The first Roosevelt administration was characterized by economic experimentation which created business uncertainty. In the second Roosevelt administration, government frustration with the disappointing recovery led to the unfortunate result of blaming and punishing private business for not investing with such measures as undistributed corporate profits taxation. There were developing technologies of that time that could have underpinned stronger investment recovery than actually occurred, but it is necessary to consider the adverse effects of antibusiness government policies to explain the poor investment performance.[6] One noticeable component of government efforts to rejuvenate growth in late 2010 is "jawboning" business to use some of a large cash hoard to expand plant and equipment spending; but this effort collides quickly with the problem that there presently is little business reason for doing this.

Inflation

The 1970s provide an example of how a failure of technology-based innovation can result in a serious inflation problem. The problem arises when the monetary authority attempts to induce growth by means of easy money. Little growth will result from such policies because the economy lacks a technology foundation for growth, at least temporarily. Inflation reflects growth in the money supply in excess of the economy's overall ability to absorb liquidity productively.

Economic growth of the 1950s and 1960s was relatively inflation-free. It reflected considerable technology push to the economy, but this impetus was largely gone by 1970. The 1970s were characterized by low growth *and* high-tending inflation, a combination which Keynesian economics deemed unlikely. An attempt to deal with inflation by means of general price controls in the first half of the decade had little effect on conditions producing inflation, and succeeded only in creating conspicuous distortions in many commodity markets, such as building materials and energy products. The decade culminated in the severe recession of 1981–1982, which was widely credited to a serious monetary policy attempt to reduce inflation.

Many economic forecasters expected more inflation than actually transpired in the 1980s on the bases of their reading of the money supply data. After all, had not easy money engendered serious inflation in the 1970s? After tightening in the effort to eliminate inflation after 1980, monetary policy was generally eased in the years following 1982, and the same period was one of *falling* inflation despite this easing. In looking back, it is curious that professional economists did not investigate the disparity in inflationary effects of easy money as between the 1970s and 1980s. During the 1970s, the economy's ability to absorb external events such as energy price shocks and internal events such as an easy money policy would have been greater had there been a technological foundation for growth to speak of. During the 1980s, such a foundation was forming based on IT; the economy's ability to absorb high liquidity productively was on the increase.

As of the present writing, anything better than weak recovery from the post-2007 recession seems less than assured. Some Governors of the Federal Reserve System reportedly are concerned that present extremely loose monetary policies create a serious threat of inflation. In spite of this concern, monetary policy has taken the direction of even more expansive policies in the form of the "quantitative easing"

(QE) program. This program aims at increasing the liquidity of an already highly liquid economy by Fed purchases of intermediate- and long-term federal debt instruments from banks. The intent of this effort is to reduce intermediate and long-term interest rates, but as of this writing, the results appear to be the opposite of what is officially intended. Official assurances that this program does not threaten serious inflation because of high unemployment have been met skeptically by financial markets, which are bemused by the inflation potential of piling liquidity atop already very high liquidity.

Composition of Demand

A technology-based boom is driven by exploitation of investment opportunities created by the new technologies, and the high returns that these investments yield depend on expanding markets for the products of new industries founded on the new technologies. All such markets eventually mature, whether they are consumer or industrial markets, or both. While consumer/industry demand for products of the new technologies is still expanding, a financial crisis may involve demand setbacks in some markets, but this will not interrupt growth in demand for products of *all* the new technologies; but a decline or disappearance of growth in *all* the markets for the products of the new technology indicates an end to the investment boom.

The boom that preceded the recession that began in 2007 was based on IT. In that recession, demand growth for *almost all* IT-related products diminished or went negative. By contrast, the recession that started in 2000 appears to be a case of a financial crisis that occurred while the IT-based upswing still had some life, for while there was a sharp downward impact on the demand for telecommunications infrastructure and hardware, the impact on demand for other technology-based goods and combinations generally consisted of no more that a growth slowdown.

The downturn that began in 1929 saw a large decrease in the sales of automobiles and a decrease in the growth rate of electric power use in general. The historical record shows that the automobile market was essentially mature after 1923, if one treats the upsurge in registrations in 1929 as a spike not indicative of a growth trend.[7] It is difficult to exaggerate the importance of automobiles in the overall industrial economy of the United States at that time in history. The extremely rapid growth in automobile demand prior to 1923 had required the simultaneous expansion of such industries as electric power production and distribution, hydraulic controls, steel, rubber, glass, and highway building. When auto demand matured, so did demand for products of these industries. Moreover, the expansion of electric power service ceased after 1930 with nearly 90 percent of rural households and establishments still unserved.

Bubbles

Economic theory has always had trouble explaining, much less predicting, bubbles because the apparently emotional drivers of bubble-like conditions lie outside the assumptions of human rationality and full information that underpin much of economic theory. Bubbles fall into two categories that are of interest for purposes of recognizing or anticipating the economic difficulties that accompany the waning of a technology-driven investment boom. The first is characterized by spending to acquire existing assets, such as gold or corn, in the belief that these can always be sold at a profit. These assets are clearly commodities and whose valuations have long been seen as volatile. Many of these trade on organized futures markets.

Bubbles of the second kind arise from situations wherein valuations of productive assets are initially driven up by fully recognizable market conditions. Inasmuch as investors cannot always perceive the exact instant when markets for products from newly acquired productive assets have become mature, at some point, the rise in valuation proceeds beyond what fundamental market conditions justify and becomes supported solely by popular belief in prolongation of the valuation increases. Examples of bubbles of the second kind include overinvestment in telecommunications infrastructure prior to 2000 and in housing prior to 2007. Both kinds of bubble end in a pattern of bidding up the prices of existing assets; they differ in how they got started. One interesting aspect of this distinction is that assets created by overinvesments in various technologies augment those posed by traditionally tradable assets such as products of mines and agriculture. This temporary addition to the slate of tradable assets plus easy money conditions increases the economy's proneness to asset price bubbles following the end of an investment boom.

At the heart of the recession of 2000–2001 were at least two bubble failures, one of each kind. One was the so-called dot.com failure, in which a number of start-up enterprises based on the Internet experienced a sharp loss of market value. In many cases, these enterprises had been founded amidst a popular wave of enthusiasm for the business possibilities of the Internet. The new firms often had little in the way of assets or income and were based on nothing more than a business plan, but in many cases, were able to sell equity shares to the public at what turned out to be inflated values. The other bubble failure occurred when it became apparent that telecommunications facilities had been massively overbuilt in relation to any but unrealistically optimistic views regarding demand for these facilities.

These developments posed a warning in two ways. First, the failures demonstrated the economy's proneness to the formation of bubbles that arise from cheap money. Second, the telecommunications overbuilding was an indicator that one important aspect of the IT boom had become grossly overinvested. Other aspects of the IT revolution experienced slowed growth, but their continued growth seemed to indicate that the IT investment boom had some remaining life. The warning of the telecom failure, in retrospect, was that if one part of the IT revolution could become invested well beyond the need for it, something similar could befall other elements of the IT revolution. This is indeed what happened, as became apparent after 2007.

The housing asset value collapse that materialized in 2007 followed the pursuit of speculative yield fostered by continuation of the same easy money conditions which had assisted the progress of the IT revolution. At the heart of the speculative aspects of the housing boom was the advancement of home purchase loans to people who were (are) unable to service these loans due to reasons of inadequate income and/or assets. This was an element in an inherent instability in financial markets described by Hyman Minsky.[8] As applied to the post-2007 recession, a key element of the instability that Minsky described can be laid at the door of the Congress, which passed the Community Development Act (which sought to encourage home ownership among low-income people), and created secondary markets for mortgage paper with an implied government guarantee: (Fannie Mae and Freddie Mac) which seemed to assure loan originators' ability to unload paper of questionable quality.

The generally high liquidity condition of the economy supported the increase in the values of housing and housing-related assets. As the situation developed, loan originators could sell loans to financial institutions. These in turn securitized the mortgage paper and sold the securitized debt instruments to investors. In an atmosphere of high liquidity, it became possible to create debt-related instruments which could be (hopefully) unloaded quickly to someone else. Each actor in the drama could assume that someone else would be left holding the bag if the structure collapsed. The structure ultimately rested on the belief of ever-increasing housing values. Thus, the speculation was at two levels: the holders of the debt paper who assumed that they could unload it quickly, and the presumption of ever-expanding home prices, which underpinned the notion that under-qualified borrowers could always rely on refinancing their houses for cash to service their mortgages. Housing asset speculation was practiced by numerous participants in the real estate markets, and the damage caused by short-term trading in these assets by a small group of professional "money men" is only a minor element in the causation of the housing market bubble.

Fraud: Increasing Frequency

Other problems grow from the pursuit of yield that persists beyond the end of a technology-based investment boom. Some of these are of psychological origin. In a time when there is a widely held illusion of prosperity and ample liquidity, people tend to be happy—and gullible. Financial frauds, of which those of the Ponzi pattern are prominent, thrive[9]; they prosper in conditions of high liquidity and a powerful seeking of high yield plus ignorance of what justifies some of the high yield rates that are offered. As a financial crisis unfolds, these structures, which depend on inflows of cash collapse and become exposed when new cash inflows fall below the redemptions rate. The Madoff affair is the recent prime example.

Many fraudulent schemes share at least one trait with bubbles: both often have their origins in some sort of legitimate activity. The namesake of pass-through frauds, Charles Ponzi, started out by arbitraging international price differences in an obscure instrument known as the *international reply coupon* (IRC). Ponzi's promises of very high returns attracted far more money than the IRC market could support, so he adopted the practice of paying earlier investors with proceeds received from recent investors and living well on his inventory of earlier cash infusions. When the scheme collapsed, many of his investors lost their savings, and his name become a dictionary entry.

For the most part, early detection of frauds has neither been a prominent objective of financial market regulation nor a very conspicuous result of such regulation. Ponzi and similar schemes promised very high rates of return to investors, and hindsight has concluded that the promise of outlandish returns should be a red flag. Madoff refrained from giving such an alarm by keeping his promised rate of return within bounds widely considered realistic.[10] Madoff attracted suspicion well before his confession, but the analyst who tried to air his suspicions was frustrated by unwillingness to hear his case on the part of regulators who should have taken interest. It appears that the private sector may be moving to fill a need that government regulation has, for various reasons, failed to provide. A number of certified public accountants have perceived fraud investigation and exposure as a logical extension of their normal business, which consists largely of filling out tax returns. This professional group has been attempting this kind of expansion for some time, but without much success. This is changing as a result of the exposure of Madoff and a number of lesser Ponzi schemes, and CPAs are receiving clientele at an accelerated pace.

Failures not Due to Criminal Activity

In not every case is a spectacular collapse due to fraudulent activity. The onset of the Great Depression saw the collapse of Samuel Insull's utility holding company structure, which rested heavily on debt. Insull was indicted for fraud, tried, and acquitted. His empire was not vulnerable due to fraud, but was simply too financially overextended to survive the crisis. Its creation was greatly facilitated by a high state of liquidity in the economy in the late 1920s. In this regard, the late 1920s are similar to the years preceding 2007. The prosecution of Insull is a prime example of politicians' imperative to blame someone, and does not encourage optimism regarding the government's ability to provide remedy for economy-wide difficulties. Politicians tend to confine their finger-pointing to small groups of people well separate from the general populace and carefully avoid blaming large population groups even when such groups may be guilty.

The recent financial collapse has had its share of failures. Several names that immediately come to mind are Bear-Stearns, American International Group, and Lehman Bros. Holdings. The first two are de facto failures even though only Lehman was the only one allowed to become formally bankrupt; the other two were technically spared from formal bankruptcy by massive government intervention and acquisition by healthier organizations. Outside the financial world, General Motors and Chrysler endured brief bankruptcies orchestrated by government. The government's motivation behind the automakers' salvage appears to be entirely politically motivated. The automakers' problems ultimately rest on worldwide excess productive capacity and a history of poor management.

The Life Cycle of a Prototype Invention

An examination of the life cycle of a technology that was an important element in an investment boom that propelled a major economic growth surge reveals several important features of how a technology cycle progresses and eventually fades. It is typical for an investment boom to be triggered by a substantial reduction in production costs accompanied by a reduction in the ability to vary products. One such technology, the laser printer, was a direct development of at least two of the major prototype inventions of the postwar era, the laser and the integrated circuit. There are many other examples of technologies whose history followed a broadly similar pattern.

The Laser Printer

The laser printer's origins can be found in the work of Xerox's PARC in the 1970s. In its earliest manifestation, the laser printer demonstrated promise for producing offset-quality print, at least in an office setting. As with many promising developments, the earliest applications were in computing centers in business establishments and universities during the 1980s. By the late 1980s, however, laser printing was becoming available in sufficiently small units to appeal to small business and some home users.

Just as the meaning of a word may best be appreciated by reflecting on its opposite, so also the progress of laser printing may be appreciated by comparison with a contemporary competing technology, inkjet printing. The inkjet printer was also an early product of the PC era, especially from the early days of the expansion of capacity of the integrated circuit. The inkjet printer received its printing instruction from the controlling PC in increments. This meant that the file being printed had to be left in computer random access memory (RAM) for the duration of the print job. Consider what this might mean for a professional tax preparer desiring to print a client's tax returns, a job that might easily take thirty minutes or more on the slow inkjet printer: this was thirty minutes not available for working up another client's return on the computer.

The early laser printers, such as Hewlett Packard's Laserjets, were provided with much larger memory capacities than contemporary inkjet models such that large files could be transferred to the printer.

This meant that, upon sending the print order, the computer was free for further work. This illustrates the effect of expanded capacity of the integrated circuit chips controlling both computer and printer. At this time (around 1989–1990), neither laser nor inkjet printing was very fast by today's standards (e.g., four pages per minute) and posed little apparent threat to commercial offset printing.

Another development from this period illustrates the limits imposed by shortages of microprocessor capacity: type fonts. Both word processor software and printers of the time lacked capacity for a wide variety of stored type fonts. This created an opening for type font hardware which came in cartridges for which laser printers had sockets. The type-face attachments were first sold by printer manufacturers, and also by third-party vendors. By the mid-1990s, however, type-face attachments had been supplanted by extensions of word-processor software, thanks to further expansion of the capacity of central processing and related chips. Also, printing speeds had increased considerably, up to twelve pages per minute and greater.

From the stand point of overall capital investment, each of these major improvements tended to trigger investment in updated equipment for offices of all sizes. This kind of upgrading was slowed by the recession following 2001, but continued to expand, if at a temporarily slower rate. This pattern has not continued in the post-2007 recession, inasmuch as sales of both computers and peripherals declined. It may be argued that this setback is due to the recession, but why did a similar absolute setback not occur in 2001? A plausible explanation is that the markets for this hardware have matured and that present-day improvements, while sometimes technologically impressive, are insufficient to justify the volume of investment needed to trigger a technology revolution and accompanying investment boom.

What would it take to revive investment in printers and their controlling computers? It may be that a hint of the future has already appeared. Commercial offset printing has one advantage: it is very economical for large production runs, such as of a magazine with national circulation or a catalog. Its disadvantage is in the form of inflexibility: even small deviation from the standard product is costly. This is a matter of potential concern in an age wherein a huge amount of customer data has been collected, whose potential in marketing is only now beginning to be exploited. However, with offset printing, it is not economical to, say, create several editions of a catalog, each targeted to consumer groups with known buying characteristics. Historically both laser and inkjet printing have offered flexibility, but neither is presently economical for high-volume work. Certain recent developments have offered the promise of substantial increases in computer-controlled printing, raising the possibility for challenging the commercial dominance of offset for big jobs. Further innovation of this technology might well create possibilities sufficiently revolutionary to induce a major surge in investment. This kind of event would be consistent with patterns of development that have long characterized the history of technology. Today's offset printing reminds of the Model T Ford, the quintessential product of mass-produced manufacturing that does not economically allow product variation. Since the day of the Model T, the trend of production processes has been the retention of mass production economies while at the same time increasing product variegation.

As of this writing, there have been improvements in the speed of inkjet printing that point toward combining the flexibility of inkjet with the speed of offset. While this line of development progresses, however, the present slump in IT and/or other investments may continue.

The Automobile

The process by which a prototype invention can enable the combination of itself and a variety of other technology threads to produce a powerful growth-inducing industrial episode is classically illustrated by the automobile. The automobile's earliest manifestation appeared about 1885. Its innovation progressed sufficiently that it became the central means of overland transportation between 1910 and 1923. Its role in the general growth that took place in that span of years in the United States and elsewhere was central because the new automobile industry became bigger than any other manufacturing industry, and rapid growth in other industries drew force from growth in the automobile industry. However, U.S. demand for automobiles was mature—at least temporarily—after 1923. The general growth induced by the rapid expansion of demand for autos previous to that year diminished after that year even though the overall condition of the economy appeared good. Then, a collapse in investment by auto-related industries was triggered by a financial crisis in 1929 and experience of the years following made it plain to industrial decisionmakers that the growth was over for a while at least.

The recent prototype invention that enabled the automobile boom was the Otto cycle engine, a unit that offered, in comparison with what went before, a compact and lightweight source of portable power. There is no better way to describe the process by which the automobile drew together different technology strains and induced their rapid growth than an illustration, which appears as Figure 4.8. In the figure, a number of principal technology inputs to the automobile appear in conventional type with dates. of approximate origin, illustrating how the prototype enables combination of a host of established technologies to produce the growth-inducing technology. A group of central activities and industries which expanded rapidly with the expansion of automobile demand is shown in bold type. It was these industries and activities whose growth collapsed with the maturity of the automobile market after 1923. If the figure resembles an explosion, the resemblance is appropriate, for that concept somewhat describes the economic growth surge that revolved around the automobile in the span of years indicated.

The auto-related growth collapse meant an investment failure that would have meant a serious recession even had the national and international financial situation been stable, which it was not. Investment failure indeed was prominent among contemporary explanations

Figure 4.8
Genesis of the Automotive-based Upswing

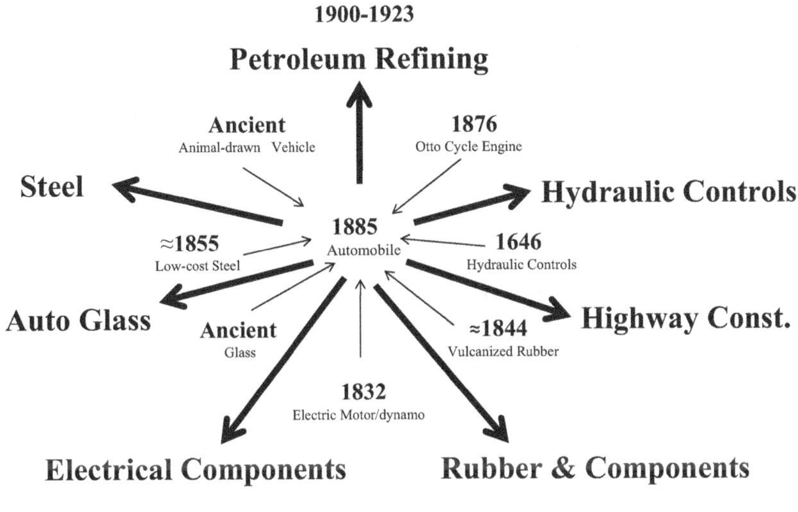

1900-1923

Petroleum Refining

Ancient
Animal-drawn Vehicle

1876
Otto Cycle Engine

Steel

Hydraulic Controls

1885
Automobile

≈1855
Low-cost Steel

1646
Hydraulic Controls

Auto Glass

Ancient
Glass

≈1844
Vulcanized Rubber

Highway Const.

1832
Electric Motor/dynamo

Electrical Components   Rubber & Components

of the event. Unfortunately, the magnitude of the financial crisis that triggered the post-1929 downturn has become the center of attention in the explanation of the great depression, and very important events in the real side of the economy have come to be overlooked. This is unfortunate, for the condition of the technology cycle makes the difference in how the economy recovers from the effects of a financial crisis. If the crisis occurs after the investment boom has exhausted the possibilities for productive capital investment enabled by the new technologies, recovery is unlikely to be quick or vigorous.[11]

While there is no question regarding the improvement potential of the modern computer, these seem likely to do no more than expand the power and speed of today's models incrementally. It is not clear that the prospective effects of such developments as optical computing, quantum computing, and even more futuristic ways of computing will have the power to trigger an impact on overall economic growth as did the electronic models that appeared in the late 1980s and early 1990s. In the classic pattern, these machines enabled combinations of themselves with other prototype technologies, such as the laser, and with established technologies.

The Modern Computer

The relation between the modern computer and growth in the economy provides an example that speaks to difficulties which the U.S. economy is experiencing at the time of this writing. The experience resembles that between the automobile and economic growth in the early twentieth century in that both experiences can be expressed in the starburst pattern of Figure 4.8: please refer to Figure 4.9. As in the automobile case, computer technology brought together a number of other technologies, some dating to as far back in history as the early days of electrical discovery in the eighteenth century. These appear in the figure in normal type. Economic activities that have burgeoned on the basis of computer technology are represented in bold type; applications of these expanded rapidly in the 1990s and into the early twenty-first-century years. In all the activities, change has been from the revolutionary to the commonplace. In other words, the power of computers to induce a powerful investment surge is largely spent; what is happening now is the replacement of the once-revolutionary technology with incrementally improved versions of itself.

Figure 4.9
Gestation of the Computer-based Upswing

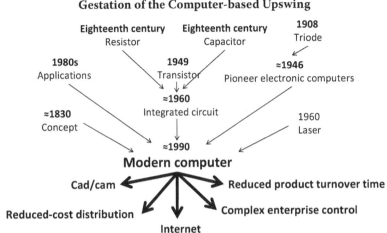

Some of the technology threads that the modern computer united were themselves recent prototype inventions in their own right. For example, the laser, which is the tool by which information stored on compact disks is input to the computer, itself was the basis for a revolutionary improvement in telecommunications. It was this telecommunications revolution that faltered following the year 2000 due to building fiber optic lines too far in advance of demand for these facilities. This overinvestment figured conspicuously among the origins of the recession of that time. Before 2000, however, the telecommunications investment surge paralleled and reinforced the computer-induced investment surge.

Notes

1. Possibly the earliest use of punch-cards to control a process was used with the Jaquard loom, circa 1800.
2. "....as a rule the new does not grow out of the old but appears alongside it and eliminates it competitively, is so to change conditions that a special process of adaptation becomes necessary." Joseph A. Schumpeter, *The Theory of Economic Development* (New Brunswick, NJ: Transactions Publishers [reprint of the 1934 edition by Harvard University Press]), 216.
3. One crucial difference between railroad investment policies in the 1950s and the 1990s was that the carriers enjoyed a level of pricing power in the latter decade that they could only dream about in the earlier. Deregulation figures heavily in the reasons behind this change in circumstances. In the

1990s, the railroads became aware that shippers were willing to pay for the improved service levels made possible by the investments; in the 1950s, the investment motives were purely cost-cutting.

4. A lack of understanding of what was causing the depression of the early 1930s led to a number of government "experiments," especially during the first Roosevelt administration. The results tended to be disappointing to the extent that the difference between "experimenting" and "fumbling" was indistinct to many observers, then and since. One serious effect of these government actions was to create enough uncertainty to discourage private sector activity that might have contributed to economic recovery. For a comprehensive development of this theme, see Amity Schlaes, *The Forgotten Man* (New York: Harper Collins, 2007), 147ff.

5. Willem Van Zandweghe, "Why Have the Dynamics of Labor Markets Changed?" *Economic Review* 95 (3rd Quarter, 2010): 8, 23.

6. For a comprehensive development of the negative effects of government policies in the 1930s, see Schlaes, *Forgotten Man* .

7. Between 1900 and 1923, U.S. automobile factory sales increased at an average annual rate of over 30 percent. If one treats the increase in 1929 as a spike, sales were effectively flat after 1923. Source: U.S. Dept. of Commerce, *Historical Statistics of the United States.*

8. Hyman P. Minsky, *Stabilizing an Unstable Economy* (New Haven, CT: Yale University Press, 1986).

9. A Ponzi scheme is one in which investors are persuaded to put money into the scheme under the pretense that the money will be used to acquire yield-bearing assets; the money is actually directly used to support a high lifestyle of the scammer, who pays investors promised high yields out of current contributions.

10. Part of Madoff's appeal was that the returns, while not huge, appeared to be consistent as between prosperous times and recessions.

11. J. R. Hicks, *Value and Capital*, 2nd ed. (Oxford: Oxford University Press, 1946), 297–99.

5

The Long Technology Cycle: Recovery

Insanity: doing the same thing over and over again and expecting different results.
—Albert Einstein

I guess in a vague sense we can say that we want energy that costs, say, a quarter of what coal electricity does and emits zero CO_2...But there are many paths to get there, each of which a realist would look at and say, "Wow ,there's a lot of difficult things along that path."
—Bill Gates

Consider a financial crisis which occurs after a technology-based boom has completely run its course, and which drives the economy into deep recession. Because governments all over the industrialized world have assumed responsibility for maintaining conditions of economic prosperity, conventional thinking calls for some sort of governmental intervention to counter the ill effects of a serious recession. Recovery from the trough that follows a downturn of the technology cycle has two dimensions: full resumption of a technology-founded upswing, and a limited recovery in which normal financial market operations are restored such that businesses can finance routine needs such as working capital requirements and capital investment in the context of normal technology change. Modern governments, especially in the United States, have tools which bear on the second of these dimensions, but they have no immediate means of inducing a technology revolution to counter a deep recession without which full recovery of vigorous general growth will not take place. The best government can do in this regard consists of *maintaining conditions favorable to pursuit of stage-one type innovation.* Absent a technology revolution, all government can do is maintain a precarious recovery that is vulnerable to further setback.

What of private business? After all, it was private business investment that propelled the boom that preceded the collapse. Unfortunately, business concentrated on investment opportunities that promised immediate high return rates and generally did not pursue what they deemed to be risky opportunities created by inventive activities, including their own. The reason is straightforward: pursuit of investment opportunities related to existing product lines tends to be perceived as more profitable than spending resources on risky stage-one innovation projects. A resumption of vigorous growth requires revolutionary technology, and private business, generally speaking, has no real incentive to call forth a technology revolution, especially one that could pose a threat to established market positions.

Government and the Recovery

The U.S. Federal Government accounts for a sufficiently large proportion of the total economy that it cannot avoid affecting the course of economic development with any action or inaction that it takes. The three policy channels that are available for dealing with serious recessionary conditions are fiscal policy (taxation and expenditures), monetary policy, and regulation. These three vary greatly in their potential for constructive result, and all offer ample possibility for destructive results. Therefore, policy implementation is highly risky and very subject to flaws in understanding of how the economy works on the part of the people charged with propounding and executing policy.

Government Fiscal Policies: Taxation

Tax reductions have proved to promote growth when they have been crafted to encourage entrepreneurial activity. The kinds of measures that have proven effective in the past include accelerated depreciation rules, investment tax credits, and measures to reduce the burden of double taxation of incomes from savings and investment. Their effect will be greatest if implementation coincides with the emergence of important new technologies from stage-one innovation. Inasmuch as this emergence can neither be predicted nor induced, tax reductions are generally *not* a "quick" fix to an ailing economy.

This has not stopped the Federal Government from attempting to deal with post-2007 recession-related problems with tax policy. The situation is this: the 2003 tax cuts were scheduled to expire at the end of 2010, and if this happens, tax rates would revert to pre-2003 levels. This would amount to a substantial general tax *increase* in an economy that remains plagued by weakness of recovery from the "great recession" of 2007–2009. Such an increase has been avoided with the eleventh-hour extension of the 2003 rates to 2012. The Administration saw the increase implied by expiration of the 2003 rates as an income *redistribution* measure, in line with its general approach to governing. However, there have been many doubts expressed on both sides of the political aisle as to the wisdom of a substantial tax increase during a time of weak economic recovery.

The trouble with temporary tax measures for investment prospects is the uncertainty they create as to future rates of taxation. This uncertainty compounds the problem of estimating the profitability of an investment even when there is a productive investment opportunity. It is true that accelerated depreciation rules and an investment tax credit enacted in the early 1960s—both temporary measures—have been credited with the growth that characterized the 1960s, but that prosperity was also aided by a substantial investment boom based on technology. One can wonder which made the growth: the temporary tax cuts or the technology-based investment boom. This question has never been investigated. It is not clear that the *temporary* extension of the 2003 rates will have any stimulative effect at all.

How can tax law change encourage recovery in the long-term technology cycle? A tax law change that favors investment income, say, by reducing the double-taxation burden on such income, is favorable to the revival of a strong technology-founded investment boom. While it may be argued that the tax cuts of 2003 were of this nature but were unfortunately timed in relation to the waning IT investment boom, such argument does not withstand close examination. The reason is that enactment of a tax measure of this magnitude is far too awkward a process for it to be used as a countercyclical tool. Moreover, even if it is so used anyway, much of its growth-promotional effect would be lost if it were generally perceived that the measure is only temporary and could be repealed in the future.[1] A tax measure designed to encourage investment and entrepreneurship can only be fully effective in achieving its object if it is perceived to be permanent.[2]

Double taxation of the income of savings, as defined in the National Income and Product Accounts, may have contributed to the debacle of 2007–2009. The housing bubble failure that triggered the recession was a long time in building. Consumers had long fallen into the practice of "saving" in the form of unrealized appreciation in the value

109

of their homes. Unrealized gains on assets are not subject to taxation, but conventional savings in the NIPA sense are taxed disproportionately heavily. Is it any wonder that consumers preferred to accumulate wealth in the form of unrealized gains? This view underscores the seriousness of the housing bubble failure and leaves a huge question as to the course of future savings. Will the stress that the 2007 recession placed on large numbers of people bring a long-term interest in saving out of current income?

If taxpayers in general perceive that taxes are likely to be revised upward, this can have a deleterious impact on resumption of technology-based growth via crowding-out financing sources that the private sector might utilize in financing the investment by which any technology surge might be exploited. This was the situation as of 2009 as a result of a huge fiscal stimulus expenditure whose positive effect on the economy was and is problematic (see below) but whose potential to create a massive increase in public debt is clear. The sharpest indicator of this condition is the U.S. Congress' apparent consideration of taxing a number of income bases heretofore untaxed. These potential tax bases include some that would have been deemed politically out of the question before the current crisis, such as some fringe benefits and mortgage interest deductibility. The goal of such tax proposals appears simply to be to increase Federal revenue. The only apparent restrictions as to what new taxes get proposed are political ones. Consideration of economic impact does not appear to play more than a minor role in the process.

Government has exerted a positive influence on technology research and development via its research efforts, and defense-related research has been prominent in this activity.[3] A permanent tax credit for private R&D expenditures would help to maintain this activity during a serious recession in which many businesses may be tempted to reduce such expenditures due to diminished revenues.

Government Fiscal Policies: Deficit Spending

The idea that an increase in government expenditures can stimulate economic recovery from a deep recession harks back to the prescription commonly attributed to J. M. Keynes. In this prescription, when private investment ceases to support economic growth, a government can replace it with its own expenditures. This idea rests on the assumption that the government expenditures induce private sector expenditures of greater amount, or has a multiplier effect. In other words, that the multiplier, μ, in a deep recession has the same force as the private investment multiplier during a technology-based expansion. This assumption is not realistic. Investment in and of itself is not a driver of economic growth, for the productivity traits of investment vary widely over the technology cycle. Investment opportunities immediately following a technology-based investment boom simply to not have productivity-raising potential remotely comparable to what they had during the investment boom. Figure 1.1 shows this in its lowest panel, in which government expenditures "G" increases with no impact on "I" or "X."

Stated differently, μ is variable in the technology cycle: high during the boom, and low to nonexistent following the boom. There are several reasons for this variability: (1) an increase in the savings rate as the downturn materializes and (2) the fact that a dollar of government expenditure under conditions of serious recession does not have the capacity to create wealth that is remotely comparable with the wealth-creating power of a dollar of private investment during a technology-based upswing. The key point with (2) is the presence or absence of a strong technology-based investment boom. Absent the technology basis of support, private investment is not likely to have any power to stimulate renewed vigorous growth, for investment demand at best will grow from the needs of normal technology growth. The Keynesian thinkers were correct in a restricted sense only: government expenditures and private investment are interchangeable in terms of multiplier impact, but the multiplier may not be much different from zero in some deep recessions.[4]

As for the increasing savings rate as a downturn materializes, one has to consider the problem of debt load, which was excessive in 1929 and 2007. In the late 1920s, the debt problem was centered on overextended stock investors, and in the recent troubles, it was centered on consumers. In a very serious recession, the instinct of debtors, both households and firms, is to use any cash flows resulting from increased government expenditures to reduce debt service expense rather than consume. This is equivalent to increasing savings and reducing consumption. In the recent episode, some of the highest-cost debt was in the form of credit card balances. During 2008, the Federal government attempted to stimulate consumption by means of an income tax rebate: checks were written directly to individuals and households. The result was that less than 20 percent of these payouts went into consumer spending. Most went into debt reduction, especially credit card debt.

The impact on the multiplier is clear, for the higher the savings rate (equivalently the lower the marginal propensity to consume), the lower is the multiplier.

The classic argument that µ disappears in the face of a government attempt to revive the economy with deficit spending is that consumers will assume that the resulting enlarged government deficits will result in substantial increases in tax rates sooner or later. This assumption generates permanently reduced consumption as consumers anticipate diminished take-home incomes. The argument goes back to David Ricardo around 1800 and assumes some foresight on the part of tax-payers. The Keynesians deal with this argument by assuming a large amount of consumer and taxpayer myopia. The protests and "tea parties" that developed in April of 2009 strongly suggest that many people are far from blind to the implications of stimulus spending for future taxation. The Ricardian argument implies that stimulus spending will not result in much economic recovery.

Bubble conditions provide an alternative (additional?) explanation for the disappearance of the multiplier in a deep recession. Recent conditions illustrate. The 2003–2007 years were characterized by high levels of expenditures financed by consumer and business debt. Will-ingness to carry debt was enhanced by the perception of rising wealth mostly in the form of rising residential real estate values. Of course, rising debt resulted in rising debt service costs relative to income, due to rising values of assets financed, not to rising interest rates, for these were very low during this entire period. Rising debt service was bearable because of the availability of even more money, such as via refinanced houses. Lenders cooperated because they shared the belief in permanently rising housing values. When housing and some other asset values began to decline, however, the burden of debt service came to be regarded very differently than when housing values were rising. Even when homeowners were able to avoid repossession, the means to service debt had to come from somewhere, and that had to be con-sumption. Widespread declines in housing values brought the prob-lems of heavy debt to the general economy in a forceful manner.

A variation on the typical classroom illustration of the Keynesian multiplier illustrates one aspect of the cyclicality of the multiplier. In the classroom example, a dollar spent becomes the recipient's income. This recipient then spends for his or her own needs, but not the whole dollar: some of that dollar is retained as saving. Let the proportion expended be represented by b. This quantity is usually presumed to be less than unity and is known as the marginal propensity to consume. If this pattern is repeated ad infinitum, then the hypothetical total effect on the economy derives from the sum of all the transactions and can be expressed as

$$\sum_{i=0}^{\infty} b_i = 1/(1-b) = \mu \qquad (5.1)$$

Inasmuch as $b < 1$, the expression $1/(1-b)$ is greater than one, suggesting that a dollar of expenditure will, through turnover, generate more than a dollar of income in the economy as a whole. This is the conventional logic for supposing that government expenditures can take up the slack created by the demise of private investment expenditures.

Expression (5.1) shows the multiplier in its simplest form. Expression (5.1) makes a number of simplifying assumptions, including no taxes on incomes and no imports. From expression (5.1), a value of b of .8 yields a multiplier of 5, which, if realistic, amounts to a huge temptation for government to engage in deficit spending, as one dollar of government spending allegedly induces five dollars worth of economic growth. What happens to this value with the addition of a little reality, such as taxes and imports? First, equation (5.1) has to be modified to

$$\mu = 1/[1 - b(1-t) + \text{mpi}], \qquad (5.2)$$

where t is taxes as a proportion of marginal income and mpi is the proportion of imports in the marginal dollar of consumption. For the value of b, use .2 instead of .8 as in common examples. This reflects the effects of a deep recession *in the presence of a very high consumer debt load.*[5] For t, use the value of .25 as an average rate of taxation on income, and for mpi, use .15. This reflects a relatively high proportion of imported goods in domestic consumption. These values yield a multiplier of 1.0, which implies no stimulative effect at all. In this example, what might have been a stimulative effect in the absence of taxes and imports has been pulled back by government as taxes, and by foreigners.

Of course, the real weakness illustrated by these hypothetical numbers is the low propensity to consume out of the marginal dollar. This result arises from consumers' high state of indebtedness. How long might the low value of b persist? The simple Keynesian answer would be that it is very temporary, and b will rise to more normal levels once indebtedness is reduced to more normal levels. This, however, presumes that consumers will quickly forget the traumatic effects of the great recession. This is unlikely, for the great recession could affect decisions regarding indebtedness for at least a generation, just as did the experience of those who lived through the great depression.

What is the experienced value of the multiplier in deep recessions? Because this value is at the center of the (popularized) Keynesian prescription for how government can respond to serious recession conditions, it has been the subject of measurement effort, but direct measurement of multipliers has proved very difficult because of the complexity of the economy and lack of historical instances of deliberate government countercyclical expenditures. Some recent estimates of the actual multiplier are attributable to Prof. Robert J. Barro of Harvard University. Professor Barro studied several periods of history in which there were dramatic increases in Federal expenditures, such as the World War II years. These have yielded a value of approximately .8 for μ under then wartime conditions. His attempt to measure μ for peacetime conditions failed to yield an estimate different from zero.[6]

Consider what this means. If μ is in the range zero to unity, the government's expenditure of a dollar yields less than a dollar's worth of employment of idle labor and capital. In this case, at least part of the government expenditures crowd out private sector expenditures by preempting financing sources, and the effect of this is to delay private sector recovery. The closer μ is to zero, the more government expenditures crowd out private expenditures, and the more prolonged will be the recession. If μ is zero, then the stimulation effort is totally futile.

If μ is positive but less than unity, there is an efficiency issue. The lower the value of μ, the less efficient will be the stimulus. If μ were equal to .1, which appears consistent with Prof. Barro's findings, then the government would have to spend \$10.00 to re-employ idle resources to the value of \$1.00. The other \$9.00 would have crowded-out private expenditures—by preempting sources of financing—and thereby delayed the revival of private investment. It is possible to get the economy back to growth with government expenditures, but the process is like unto setting out on a long trip in a car that has only low gear: you might arrive at the desired destination eventually, but only after prodigal expenditures of time and fuel. Proponents of government stimulus spending, when reminded of the apparent failure of the Japanese government to revive the Japanese economy during the

1990s with spending, tend to claim that the Japanese stimulus effort was too small and too belated. The U.S. stimulus expenditures of 2009 do promise to be bigger and timelier than those of Japan of the 1990s. As of the time of this writing (late 2010), the beneficial effect of the stimulus program in the United States is virtually undetectable. Futility is futility regardless of the scale.

The Current Fiscal Stimulus Experience

As already mentioned, 2008's income tax cash rebate program was unsuccessful at generating consumer expenditure increases because most of the money was used to pay down debt. In March of 2009, federal stimulus expenditures of $862 billion were authorized by Congress. Although senior members of the Obama administration have recently extolled the success of this program, it is not immediately obvious that there has been any stimulative effect. The overall unemployment rate, as of November 2010, is only slightly below what it was at the official trough of the recession in 2009. Moreover, of the unemployed, the proportion that has been out of work for more than twenty-seven weeks is at a post-World War II high.[7] As has already been observed in this book, the results of the IT revolution include a changed productive paradigm and, consequently, a different labor market. Many of the long-term unemployed were in the professional or middle management ranks prior to their loss of jobs. The essence of the technology revolution is that employers are enabled to achieve profitability goals with fewer hands. In some cases, those laid off have become obsolete, but all have become redundant. This suggests a high structural component in the official unemployment rate that is unlikely to respond to measures that at best are effective with purely cyclical unemployment. A high percentage of long-term unemployed in the official unemployment rate is not the only hint of the structural character of current unemployment. If the number of people who have dropped out of the labor force due to discouragement, the number who want a job but are not actively searching for one for reasons other than discouragement, and the number of underemployed workers are added to the number of officially unemployed, the total has been estimated at 17 percent of the labor force. This number may be a better measure of the extent of the recession's aftereffects than the official unemployment rate.

What the administration is claiming is in terms of jobs saved—those that were not destroyed by the recession. This number is based on economic modeling in which the predicted job losses absent the stimulus are compared with actual job losses. Not surprisingly, jobs saved is a controversial number. There is some basis for the claim in that federal direct aid to states has prevented the loss of teacher, police, and fire-fighting positions, but beyond that, there is little evidence that *private* sector hiring has been stimulated at all. This is significant, for as between the public and private sectors, the latter has a monopoly on growth-generating power.

As in the 1930s, the stimulus of 2009 was "sold" with the proposals to spend on physical infrastructure projects. Apart from the logic that stimulative effects of infrastructure projects are temporary in nature, as of the late summer of 2010, only about one-third of the $230 billion designated for infrastructure projects has actually been spent. Other parts of the spending program may have some kind of merit, but make little sense as short-term measures to offset the recession. For example, energy-related expenditures have been devoted to "clean energy" projects, including the financing of start-up companies that satisfy this definition. This is not a short-term stimulus, if it has any stimulating effect at all. It appears to be an effort to induce a technology revolution, an unlikely prospect, at least in the short run.

Can the Federal Reserve Prevent a Collapse?

Conditions following the peak of a technology-based boom typically include a high degree of liquidity in the economy. This encourages the formation of asset price bubbles and is favorable to fraudulent schemes and other possible attractors of money seeking high yields. The problem is that in the waning upswing, truly productive high-yield investment opportunities become increasingly hard to find, so that money chases other opportunities of a less-desirable kind. Ideally, monetary policy under such conditions should tighten, but tightening once an asset price bubble has become widely perceived would be likely to produce the feared setback that inspired the easy money policy in the first place. The tightening action has to come *before* the onset of asset bubbles. Achieving this is no mean challenge, for it requires recognition of the maturing of the technology cycle.

Generally, the easy money policy has been justified by a desire to maintain conditions of low unemployment and to promote business activity. Tightening of monetary policy in order to diminish the likelihood of the formation of asset bubbles has been seen as discouraging business activity. Under conditions of a waning technology-based investment boom, however, the conflict between a strategy of forestalling bubbles and one promoting business activity is less important than it might appear on the surface. If business' most productive investment opportunities are disappearing, then the kinds of business activities that easy money policies encourage will increasingly take undesirable forms, including investments that rest on unrealistic assumptions of permanently rising asset valuations. Consequently, there is little to lose and much to be gained from concentrating monetary policy on discouragement of the formation of asset price bubbles.

In 1928, the New York Federal Bank under the leadership of Benjamin Strong, was worried about the apparent growing stock market boom, and advocated tighter monetary policies to head off a bubble. At the same time, the Fed's Board of Governors in Washington feared any deviation from an easy money policy aimed at promoting business activity. This conflict yielded a policy too weak to head off the bubble, but "...too restrictive to foster vigorous business expansion."[8] The present author would argue that under the-then extant real economy conditions, no amount of monetary ease would have been sufficient to foster vigorous business expansion because the productive investment opportunities that grew out the automobile- and electric power-centered boom had been temporarily exhausted; consequently there was no gain from even attempting to stimulate growth. Heading off bubbles should have been the sole object of monetary policy.

A seemingly similar conflict affected monetary policy following the recession of 2000. This recession was proximately triggered by failure of the dot.com bubble, and this drew most of the policy attention. However, the recession also involved failure of another bubble: investment in telecommunications facilities. This investment failure aspect of the recession may not have received the policy attention it deserved, for the monetary response, which appeared to take inspiration from the response to the stock market correction of October 17, 1987, consisted of pumping a large amount of liquidity into the economy. There had been little in the way of tightening in the several years prior to 2007 that might have countered the gathering bubble in the values of housing-related assets. The impact of monetary ease on the rate of productive investment prior to 2007 was probably nil, but its impact on bubbles in tradable commodities became sadly obvious.

Figure 5.1 illustrates an appropriate policy response as a technology-based upswing approaches its end. In the absence of any action

Figure 5.1
Technology Cycle Upswing End Game

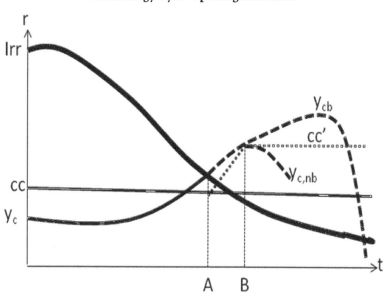

affecting the cost of capital (CC), money-seeking yield will switch from productive investment (Irr) to investment in tradable assets such as commodities or housing-related paper (represented by the dotted portion of the y curve (y_{cb}). Inasmuch as this kind of "investment" takes its justification solely from the belief in continued appreciation in the price of the asset(s) in question, the possibility of an asset bubble is very real. The figure shows this kind of return rising for a while, and then collapsing. An increase in the interest rate structure (CC') from pt. A to pt. B would take much of the romance out of speculation. It is doubtful if this kind of maneuver has ever been carried out successfully, the primary reason being the monetary authority's bemusement with the idea that continued monetary ease can stimulate business activity. It cannot when business' productive investment opportunities have temporarily disappeared. If the Federal Reserve were to recognize this, then the timing of the maneuver would be critical. Forestalling the housing bubble that began to collapse in 2007 would have avoided the creation of a lot of housing-related paper that went worthless with the decline in housing values and caused so much trouble for financial markets in general. Waiting for the crisis to become fully apparent before increasing rates invites several consequences that are likely to impair the recovery. These include a substantial decline in the capital value of bonds, including Treasuries, which would be a shock to the recovery. Given the difficulties of recognizing the precise instant at which an increase in rates is appropriate, it is better to increase rates too soon than too late. A "too soon" increase would be in the context of a still-active investment boom in which the economy could absorb the impact of the increase with minimal damage.

At this point a comment on the Federal Reserve's policy proposals as of late 2010 is in order. The Fed appears to have convinced itself that high unemployment is a purely cyclical, not a structural phenomenon, and that there is no danger of inflation. Therefore it seems to see no difficulty in a massive money-printing exercise, for this is the implication of the Fed's large-scale purchase of longer-term Treasury securities held by member banks of the Federal Reserve system. Such a policy will *not* help the unemployment problem, for much of it is in all probability structural, not cyclical. How is it structural? Corporate profits are holding up surprisingly well. Moreover, corporations have discovered that they can operate successfully with reduced employment. Many of the layoffs of professional and middle management workers were of people who became obsolete or in some sense redundant as a consequence of the preceding technology-based investment boom. Therefore, where do employers discover an incentive to hire back people whom they no longer need?

The most likely result of a money-printing surge is explosive inflation. Why is this? Consider the issue from the viewpoint of investors holding inventories of U.S. debt. The announcement of quantitative easing (QE) produces the instantaneous thought of future inflation. That prospect creates a strong incentive to sell, and as a result, the Fed finds itself buying in a market that is experiencing massive selling pressure. Unfortunately, the amount the Fed proposes to spend in QE is far less than the possible offering by bond sellers. The potential is for rising (not falling) longer-term interest rates and rising inflation. QE has been defended as noninflationary because of the high unemployment rate. However, this assurance is based on the assumption that unemployment is simply cyclical, which is unlikely. High unemployment and simultaneous rising inflation will not happen according to conventional economic theory, but the Fed may have discovered how to make it happen anyway.

Repair of the Financial System

If government expenditures are not going to be effective at reigniting the upswing in the temporary absence of a technology-based push, then is there anything that government can do in the short term? The answer is easily stated: restore financial markets to a normal level of operation. This is a necessary (but not sufficient) condition for the restoration of vigorous growth. Restoring the financial system to proper working order will not guarantee resumption of boom-like economic growth, for only a renewed technology push can accomplish that. The essential role of financial markets is as an intermediary between savers and entrepreneurs whose activities create productive wealth. Any impairment of the financial system's ability to carry out this role is of itself a threat to the possibility of a strong recovery at any time.

Restoration of the financial system is necessary in order that businesses of any size be able to have access to credit to enable day-to-day operations. This includes such basic short-term needs as access to working capital financing.[9] While the technology foundation for significant renewed economic growth may be temporarily lacking, the economy can function at tolerable levels of unemployment when it is possible to finance investment to achieve incremental improvements in existing technology. Moreover, serious impairment of business' access to capital, bank or otherwise, is an impediment to financing growth once the technology basis for it materializes. When financial markets cease to function, the economy is unable to achieve even a modest growth potential.

The onset of the financial crisis shifted the emphasis of Federal Reserve action from originator of monetary policy to its role as bank regulator. The Fed shares its bank regulatory role with other organizations, such as the Federal Deposit Insurance Corporation (FDIC). The initial regulatory approach to the crisis was influenced by the Japanese experience of the 1990s. In this, banks' lending ability was impaired by their deteriorated capital condition. The Japanese economy suffered under recessionary conditions for most of that decade, and did not resume growth until the government took positive action to clear the toxic assets from bank balance sheets. The thinking evidently was that once the banks were again able to lend, then other government measures, such as expansionary monetary policy and government expenditures, would become effective at fomenting growth.

Inasmuch as the recent U.S. downturn was triggered by reduction of the value of housing market-related assets on bank balance sheets, the easily stated solution is for the Treasury or the Federal Reserve to acquire the "toxic" assets and remove them from lender balance sheets. Unfortunately, stating the solution generally proves to be far easier than executing it. The problem lies in the pricing of the so-called toxic paper on lender balance sheets. Current market valuations, effectively zero, are not very helpful as guides as to what to pay, for "relief" at such valuations would force recognition of the zero valuations on their books and thereby trigger insolvency of many institutions. To impute a higher valuation on problem assets, as might be justified by a projection of values given a recovering economy assumption, would create a political exposure in which government officials would be accused of a giveaway of taxpayer money in order to salvage the interests of bank managers who were probably complicit in bringing on the crisis in the first place.

The problem of valuing the real estate-related paper on bank and other lender balance sheets arises at least partly from the interpretation of the Fair Value rules of the Securities and Exchange Commission that would require banks and other holders of certain real estate-related paper to mark the value of these instruments to current market value.[10] Inasmuch as the market for a lot of this paper is dormant, this value is zero, at least in the very short run. This is in spite of the fact that some of these assets still produce income flows for their holders. When there is an income, an alternative valuation method is available, namely the discounted value of the expected income stream. This valuation method would relieve considerable pressure on the balance sheets of many institutions and thereby avoid problems of valuations under distressed market conditions.[11]

It is clear that the manner in which the so-called toxic assets are valued, affects the financial condition of lenders, and therefore the perceived magnitude of the crisis. A literal interpretation of the mark-to-market rule will result in a larger number of insolvent lenders than another valuation method that recognizes that many of the mortgages underlying the toxic assets are still servicing their mortgages despite the condition that their houses are worth less than the value of their mortgages. William Isaac, a former FDIC chair, has placed much of the blame for the crisis of subprime mortgages on the requirement that banks mark their housing-related paper to current market value, effectively zero. Such revaluation would have widespread negative effects. These include directly reducing apparent bank capital below levels necessary to comply with solvency regulations, and indirectly by triggering margin calls from other lenders for loans for which the residential paper has been pledged as collateral.[12] Any regulatory rule that has the effect of increasing the number of insolvent lenders threatens the larger economy by its effect of diminishing lending, on which countless smaller businesses depend for financing their working capital needs. This opposes the supposed overall goal of restoring financial markets to normal operation. Indeed, one of the announced goals of QE is to increase banks' money reserves to the point where they can write off some or all of their "toxic" real estate paper without threat of insolvency.

The recent action by the Financial Accounting Standards Board (FASB) to ease the mark-to-market rules as related to markets that have become nonfunctional should relieve banks and other lenders by making toxic assets less so.[13] The valuations of such assets will be less of a threat to lenders' ability to maintain required capital standards. This is unlikely to be a complete cure-all for the financial market. For institutions that are effectively insolvent in spite of increased valuation leeway for nonperforming assets, the bankruptcy process should be considered. This recalls the Japanese experience of the 1990s, wherein the economy failed to recover fully until the banks' bad assets had been largely removed. Bankruptcy law contains many provisions by which many of the affected institutions could be reorganized successfully.

The existence of "toxic" assets that still produce income gives some justification to the Treasury's March 2009 proposal to encourage private sector participation in buying toxic assets from banks. This proposal supposes that there are riskophilic investors who will bid enough for this paper to interest banks in selling. The Treasury's proposal places most, but not all, of the risk on the government. It remains to be seen how much private interest this scheme will elicit after the recent FASB rule-change proposal,[14] whose effect will be to strengthen many lenders' bargaining position relative to would-buyers of housing-related assets.[15]

Regulation

It is the normal pattern for Congress to enact legislation whose wording fails to disclose important guidance regarding how the objectives of the legislation are to be accomplished; indeed, a certain amount of vagueness on this matter is often politically necessary to secure majority support for the measure. The details of enforcement are left to this or that government agency, and these propound the rules by which the legislation becomes implemented. This is the source of a huge amount of government regulation. It is inevitable that the regulatory process is profoundly influenced by the interests of those economic activities that are affected by the regulation. This whole process gives regulation in general a tendency to respond to problems recently recognized and imparts no impetus toward anticipating future problems. The mark-to-market episode shows how a rule that seems to improve market functioning in one set of circumstances can hinder efforts to deal with unforeseen new circumstances. There is no reason to suppose that this will be the last instance wherein regulations stumble over one another to confound efforts to deal with some future problems.

When there is a serious financial crisis that precipitates a downturn, attention naturally turns to the idea that weakness in the structure of government regulation is at fault. While regulation can always stand to be improved, four things should be kept in mind. First, regulatory revision will have little impact on the immediate crisis; its effects will be felt mostly in the *next* financial market crisis. Second, new regulatory regimes are as subject to the law of unintended consequences as any other action of the government that affects the economy. *To wit*, the literal interpretation of the mark-to-market rule had the effect of magnifying the apparent extent of bank insolvency in 2007–2008. Third, while the connection between regulation and recovery of general growth founded on a technology boom is extremely indirect, regulation has the potential to inhibit the development of such a boom. Fourth, government offices established to carry out one or another regulatory intent of Congress are traditionally staffed inadequately to carry out their assignments. Congress wishes to avoid creating of an open-end cost center, and therefore imposes a budget constraint on a regulatory body. One of the ways in which a regulated activity can respond to regulation is to circumvent it, and it can be powerfully ingenious in doing this. No better example of a result of this condition can be found than the regulatory toleration of the erosion of lending standards which played a major role in the failure of the housing bubble in 2007.

Perhaps the greatest threat to the possibility of a renewed growth surge based on new technology in the still-unfolding aftermath of the recession of 2007 is the idea that some companies are "too big to fail."

123

The problem that this threatens is that in the future, large companies that have failed the test of the marketplace will be kept alive at taxpayer expense on the pretext that their failure would be "too disruptive." This disruption would fall heavily on employees, perhaps unionized, and retirees of the companies in question. Consequently, the fear of disruption can be traced largely to politicians' fear of loss of the votes of large numbers of affected workers. The expense of keeping effectively moribund companies alive diverts financial resources that might otherwise be available to finance a real growth recovery.

Of course, it may be argued that government control of an otherwise moribund company can be a restorer of that company's efficiency and competitiveness. This is unlikely, however, unless there are no important voting blocs that are complicit in the company's failure prior to government control. If this condition is not met, then no great changes will occur, and the companies' decline will only have been delayed by the presence of government in company strategic decision making. Experience with government-controlled companies is not encouraging. A prime example is British Leyland, which under government/union control has disappeared, its assets having been sold. A second example is Petroleos Mexicanos, which for many years has made sense only if viewed as a jobs program, not a commercial enterprise.

General Motors and Chrysler quickly come to mind as companies faced with a future of decline. Union influence in both cases is likely to inhibit work rule changes necessary to adapt operations to the advantages conferred by changing technology.[16] It presently appears as though government influence will intrude on basic management decisions, such as what to produce.[17] A dramatic tightening of CAFE standards with present technology implies an increase in the proportion of small cars in the product mix. Needless to say, the car-buying public was not consulted in this matter, but if car buyers refuse to endorse this prospective future product line, then the carmakers will collapse in spite of the government's efforts. It is natural to ask why the taxpayers at large have to bear the cost of delaying an inevitable collapse.

The 2007-present downturn has induced some discussion of financial market regulation as an approach to heading off asset bubbles before they get started. Interstate comparisons have lent some fuel to the idea that regulation can be used in this manner. Vermont had in place restrictions on banking activity that forced banks to maintain relatively stringent standards regarding approvals of mortgage loans (including requirements for demonstrated ability to make down payments and to service mortgages). Partly due to these measures, Vermont for the most part had neither a boom nor a collapse of its housing markets as did other states, such as Nevada and California. This kind of comparison gives some plausibility to the idea of stringent regulation of lending standards nationally. Not surprisingly, lenders and homebuilders are violently opposed to this idea. These argue that the result of such regulation would be the elimination of a powerful source of economic growth nationwide. However, inasmuch as infamously loose lending standards are at the heart of speculative demand for housing it is difficult to see what harm more stringent standards would have.

While this kind of regulatory approach would have helped forestall the housing bubble, it is difficult to envision how application of regulation to trading of other kinds of tradable assets might be applied in other markets without doing irreparable harm. One danger is the impairment of the ability to hedge for the purpose of locking in a future price. This is a common strategy used by farmers who must plant a crop months before the crop is sold. In a hedge, a farmer is long in the physical commodity (e.g., wheat, or corn) and takes a short position in the futures markets. A decline in the price of the physical crop is offset by a gain in the futures position, and vice versa. For a hedge to work, however, there must be a speculator at the other side of the bargain who is willing to bet opposite to the farmer. Therefore, it is hard to imagine restriction on speculation that would not interfere with legitimate hedging operations.

Government and Innovation

One of the important roles that the Federal Government can play in fomenting recovery from a deep recession is that which it has played for the entire history of the republic: a large buyer of novel technology and research and development relating to new technologies at the stage-one innovation level. This activity assumes risk at a level that is likely to discourage private sector investors. National defense-related research has been a rich vein of new technology which the private sector has been able to develop to the benefit of the entire economy. The Internet and global positioning are but two examples. Other government technology initiatives flow from agencies such as National Institutes of Health, National Science Foundation, Department of Energy, and National Aeronautics and Space Administration. Bringing new technology from the laboratory curiosity stage to that of being a useful product is the riskiest part of the invention–innovation process, and is the one least likely to attract interest from a private sector, especially one suffering from the after-effects of a financial market crisis. Assumption of this risk by government makes a solid contribution not only to short-term cyclical recovery, but to eventual resumption of strong growth.

It does not go too far to observe that the most successful government and government-sponsored research and development efforts have been largely away from the public eye. Being out of the public eye often means being out of the eye of the Congress, and therefore not subject to the influence of narrow commercial and/or parochial interests. Projects meeting this condition can be pursued objectively until they either yield something useful or are abandoned when they appear to be dead ends. Not surprisingly, many of the more spectacular successes of past government R&D have originated from national defense-related research, and these efforts have become formalized in one agency: Defense Advanced Research Projects Agency (DARPA).

It is not surprising that DARPA's successes have inspired emulation, and one of the more recent manifestations of this is Energy Advanced Research Projects Agency (EARPA) in the Department of Energy. On the whole, it is probably a good idea to adapt the DARPA pattern to energy research, but there is a potential problem that arises with research in an area that has become thoroughly politicized, for this opens at least some of EARPA's work to powerful outside political influences which may or may not work to encourage objective research. To illustrate this point, consider fuel alcohol. When fuel alcohol is based on corn, it consumes more fossil fuels in its production than the energy content it delivers, but a commercially questionable but very substantial fuel alcohol industry is vigorously defended by a coalition of corn-related interests. Therefore, there is a large barrier to terminating government subsidy support. However, EARPA itself may prove useful when its projects are less in the public eye than the fuel ethanol program. Fuel alcohol is but one possible area of energy-related research, and there are plenty of opportunities for EARPA to operate quietly.

One of the necessary requirements for the start of a technology upswing is an entrepreneurially friendly tax regime. Entrepreneurs' activities drive stage-two innovation processes. There is an unfortunate political thread in the United States that is based on envy which calls for heavy taxation of high incomes. People who support using tax policy as a redistributive tool often think in terms of well-publicized accounts of huge compensations to chief executives who are perceived as looters of their companies and "coupon clippers,"—heirs to old fortunes. The political urge toward "soak the rich" tax measures tends to be especially strong following a financial market collapse and produces the naming and punishment of scapegoats: for example, Samuel Insull in the 1930s and some senior Wall St. and Corporate figures recently. State–state comparisons also illustrate the downside of treating the taxing power of governments as an income redistributive tool. California is a relatively high-tax state that fell into fiscal difficulties in the great recession. A substantial part of its tax revenues have come from high-income taxpayers. It happens that California has suffered a loss of a million citizens in each of the most recent Census decades following a long history of population growth. Could it be that heavy and worsening tax burdens plus anticipation of even worsening conditions have made California seem less like the Garden of Eden that it used to seem?

While there may be some merit in popular perceptions of rich taxpayers, use of tax policy as a redistributive tool also penalizes those who have gained wealth through their own creativity and energy. These qualities and efforts result in the creation of wealth which did not exist prior to entrepreneurial effort and whose benefits are widely distributed in the economy. Indeed, the prosperity generated by the stage-two innovation of a group of prototype inventions/discoveries reflects massed creativity. It is extremely important that countercyclical fiscal policy not include any but light taxation of capital-derived incomes.

Can Industrial Policy Bring Recovery?

The term "industrial policy" is usually taken to mean a government policy of influencing the flow of capital into particular industries in order to further some public policy goal. The choice of industries to favor can depend on any of a variety of political or economic reasons, but many become publically justified on the grounds of claims of benefits related to economic growth. U.S. industrial policy in 2010 consists mostly of a mix of subsidies, import restrictions, and other measures intended to promote environmental goals. Some of these measures were justified as economic stimulus at first, but when their irrelevance as short-term stimuli became widely obvious, the justifications changed to "investments" affecting the government's desired direction of future growth. While the term "investment" is used to convey the idea of building a basis for future growth, these policies illustrate the point that there is an enormous variation in the quality and effectiveness of actions that can in some sense be termed as "investments."

Brief History of U.S. Industrial Policy

The United States has a long history of government encouragement of particular industries. In the nineteenth century, government financial aid was used to support infrastructure development such as canals and railroads. A more recent example was the Interstate Highway project which originated in the late 1950s. Federal support has also gone to the fostering of commercial airlines through the subsidizing of airport construction and the system of air traffic control, and commercial inland waterway development, such as dams designed to maintain minimum channel depths for navigation.

The forms of government support have varied across a gamut including grants of land, loan guarantees, tariffs on imports, tax credits, and others. Often, some payback of the costs of government support of benefitted industries has been required, but this can take various forms other than a narrowly defined repayment, as with a loan. To be successful, infrastructure assets acquired with the aid of industrial policies have to result in economic growth. The idea is that the growth generated by the industrial policy will be a basis for an enlarged tax base such that taxpayers are recompensed indirectly for the direct costs of the industrial policies.

For example, land and cash grants to railroads in the nineteenth century typically carried a requirement of ceilings to freight rates on government traffic. While these ceilings were usually set high enough that they did not pose a revenue problem for the railway companies at the time they were built, they were never adjusted for inflation, and saved the government huge amounts of money in shipping costs (relative to commercial rates) in later episodes of high government shipping activity, such as World War II. It has been widely accepted that the economic development that accompanied or followed expansion of the railway net generated more than adequate recompense for the early subsidies.

Costs of maintaining navigability of the inland waterway system were borne almost exclusively by the government until a user charge in the form of a diesel fuel tax was imposed in the 1970s. There is no realistic claim that a fuel tax has repaid the historical costs of maintaining the waterway system. Rather, continuation of the subsidies is repeatedly justified on the grounds that bulk commodity transportation on inland waterways is cheap and results in copious desirable multiplier effects in the form of industrial development. The Tennessee Valley Authority (TVA) project is frequently held up as a positive example. However, the TVA development of the Tennessee River system had multiple purposes, and arguments by waterways interests face the problem of isolating the benefits of navigation from other benefits, such as electrification of a large region of the country and flood control in a valley prone to flooding. Perhaps a better example comes from the Ten-Tom project of the 1970s. This project has connected the Tennessee River with the Tombigbee River for navigation, thus creating a navigable shortcut connection between the Tennessee River and the Gulf of Mexico. Proponents of the Ten-Tom held out the prospect of a massive industrial development in Eastern Mississippi, but this has yet to happen. This disappointment should have been foreseeable at the time, for *Western* Mississippi has always had available the lowest-cost navigable waterway of them all—the lower Mississippi River—and was and is regarded as underdeveloped industrially.

The pattern of industrial policies in the late twentieth century has moved away from the narrow idea of providing transportation infrastructure that dominated the process in the nineteenth century, and in the direction of encouraging industries that are based on technologies that have not developed beyond stage-one innovation. This change is important because it burdens industrial policy with the risk that that the supported technologies' emergence into stage-two innovation may be long in the future, or never.

During the 1970s, federal funding for industrial policies tended to be justified by a need to achieve something called energy independence. This was in response to a dramatic increase in the costs of energy products to Americans and an increasing dependence on imported energy supplies. There is no better example of this than the Great Plains Gasification project. This was an attempt to convert low-grade coal mined in the Dakotas into pipeline-quality fuel gas (methane) that could be moved to market efficiently. This project emerged during a period in which it was believed in government circles that the supply of domestic natural gas was in terminal decline. The plant was built for a cost of about $2 billion with the support of a consortium of private investors, but the private sector debt used to finance the plant was government-guaranteed. When the perception of terminal depletion of conventional gas was discovered to have been little more than a market reaction to government price regulation, the project was abandoned in the mid-1980s when the price of gas collapsed.[18]

As of 2010, several issues dominate government subsidizations of industries. The clean air issue has been a driver of subsidies since the early 1970s. The list of target pollutants of the 1970s, such as sulfur dioxide, has been augmented by greenhouse gases, of which CO_2 is the largest-volume example. A revival of the price of petroleum products and natural gas has brought renewed interest in the energy independence problem. Because much progress was made on the problem of pollutants as recognized in the 1970s, today's industrial policy emphasis has shifted to the greenhouse gases. The supposition here is that these gases are aggravating a long-term trend toward warming of global climate. Global warming is a long-term apparent trend with several centuries' history. The question of whether or not human activity plays a considerable role in furthering this trend remains controversial.

Industrial Policy and Stage-one Innovation

During the last half of the twentieth century, public policy goals justifying industrial policies began to involve technologies still in stage-one innovation. This means that many of the technologies benefitting from government subsidies are intended to supplant older technologies with technologies having higher measured costs than the technologies they are supposed to replace. Closures of the unfavorable cost gaps are dependent on technology breakthroughs whose timing is totally uncertain. Without the breakthroughs, the only way the promoted technologies will be adopted and expanded by the private sector is an ongoing stream of government largesse large enough to overcome cost disadvantages. Supporters of the subsidized stage-one technologies argue that a part of the cost disadvantages are only perceived because prices presently being paid by consumers for, say, conventional electric power, do not cover the entire costs of its production, especially costs imposed by air pollution, climate deterioration, and other effects that do not pass through market transactions. Various measures that would impute a market value to pollution have been proposed, such as cap and trade schemes and carbon emission taxes. Enactment of measures that would impute a cost of polluting to emitters of greenhouse gases has so far foundered on basically political grounds—such measures would noticeably increase the costs of consuming products of greenhouse gas-emitting industries, and this could be detrimental to re-election chances of supporters of such measures.

The burden of lack of a cost breakthrough diminishes the chance that an industrial policy will achieve its stated growth goals through support of stage-one industries. A comparison between subsidizing industries still in stage-one innovation and those that are already commercially viable shows why. In the former case, any scheme that *realistically* imputes a market value to emissions of undesirable by-products will impute a value to the *combination* of costs attributable to undesirable emissions, such as greenhouse gases, and those of production of the desired output, such as electric power, in excess of costs that would prevail if the underlying technology were to undergo a cost-lowering breakthrough sufficient to move it into stage-two innovation. If the subsidization goes to an industry that is already commercially viable—i.e., beyond its stage-one innovation, then the cost imputation reflects purely the cost of the pollution without the burden of the lack of the breakthrough necessary to make the technology commercially viable.

The success of an industrial policy hinges on the government's willingness to terminate the subsidy support when it becomes apparent that the benefitted industry has little or no chance of becoming commercially self-supporting. Termination of the Great Plains Gasification project offers some encouragement in this regard. Unfortunately, any government-subsidized project tends to acquire a political constituency that will resist project termination almost from its start, and the longer the subsidies keep the project alive, the more entrenched the defending constituencies become. This spotlights the risk inherent in an industrial policy that extends support to industries that work with stage-one technology, for in the absence of the breakthrough that propels such technologies into stage-two innovation, the time when benefitted industries can be weaned from subsidies is totally uncertain and may be in a distant future.

There is no better example of this than the government's efforts to foster the fuel ethanol industry. Subsidies include a 55 cent/gallon import duty on imported sugar-based ethanol, tax credits for corn ethanol producers, and mandatory admixtures of ethanol in motor gasoline. These measures have been justified by the claim that they promote energy independence, as ethanol in motor fuels supposedly displaces gasoline made from imported crude oil. Studies, however, have concluded that corn-based fuel ethanol may consume more energy in its total production than the energy value of the produced ethanol. Yet the corn interests are vigorously campaigning for continuation of the subsidies. As long as these interests are successful, the programs will continue, possibly forever. One idea that defenders of fuel alcohol subsidies use is that the corn ethanol experience will contribute to the development of a truly economical way to produce ethanol from biomaterials now considered waste products, such as switch grass or corn stover. Again, when might this happen? There is no guarantee of soon.[19]

The political strength of the corn ethanol and other "green" energy industries may diminish for a reason that would have astounded promoters of the Great Plains Gasification Project: the discovery of huge quantities of natural gas in continental North America. This has been a result of the oil & gas industry's development of the ability to produce gas economically from shale formations that were deemed to be uneconomical sources in the 1970s. Needless to say, this development has occurred in a deregulated market. Now the prospect is for domestic gas supplies sufficient to support many decades of U.S. consumption, and there is now talk of the United States as a net gas exporter.

Industrial policies have been tried in a number of countries. Their records, as fomenters of growth, have been spotty at best. Administrations of industrial policies have been assigned to formal ministries in several instances, with Japan's Ministry of Trade and Industries as a prime example. Formal bureaucratic industrial policy administration has commonly fallen into a pattern of backward horizon. This is because administrators are human and have no effective view of the technology future. Consequently, the recommendations for encouragement involve established industries, or, at best, industries thought to have a future whose technology has yet to emerge from stage-one innovation.

Industrial Policy and Technology Revolutions

The often-stated object of an industrial policy is to channel capital into industries with high potential for growth. The insurmountable problem in doing this is the inability to identify growth industries in an era of technology flux. Most of the time, this problem does not receive attention. Even when there is an explicit desire to consider possibilities of the technology future, the available choices of which industries to favor remain confined to the known. What passes for anticipating future technology amounts to playing roulette with which combinations of known technology threads in stage-one innovation will experience critical cost-reducing breakthroughs. Neither industrial policy agencies nor any other government body has the power to do more than throw money at this or that research and development effort, and this is far from inducing a desired breakthrough.

Provision of infrastructure in geographic regions that lack such facilities using well-known technology—the nineteenth-century model—is a way to avoid the problem of foreseeing technology future. This is not an effective option in the United States today because the growth impact potential has largely been exhausted: the United States is already well-provided with transportation infrastructure. Somehow, catching up on deferred maintenance of a bridge here or there does not have growth impact comparable to building a bridge where none previously existed.

One of the ways by which both government and private sector leaders attempt to deal with the technology future is by anticipating future cost reduction achievable with large-scale production. This kind of thinking has been common in the so-called high-technology industries, wherein products have been priced at a loss but in anticipation of future economies of large-scale production runs. Something like this thinking is taking place presently with regard to electric cars. Present designs of these vehicles all suffer from the very low-power density of even the most advanced batteries, which gives these vehicles low capability and high cost. The thinking evidently is that large production runs will reduce battery cost sufficiently below today's elevated levels that electric cars will become attractive to car buyers. However, in past instances in which there has a combined learning curve and scale-based reduction in costs, the process has also been supported by a basic cost breakthrough. Battery technology still lacks this.

The Riskiness of Predicting the Future Course of Technology

The problem of longer-term anticipation of a technology revolution, crucial to a successful industrial policy, can be illustrated by means of a mental experiment. Consider the subset of the technologies related to electric power production and distribution that are presently and recently in stage-one innovation. As noted above, for a technology in stage-one innovation, a necessary condition is a dramatic diminution in the costs of its application. It is possible to make reasonable conjecture regarding the impact of such a diminution, but that event is of highly uncertain timing. However, the timing of breakthroughs capable of propelling a technology thread from stage-one to stage-two innovation is crucial in the matter of how overall economic growth proceeds. Moreover, the *order* in which breakthroughs occur can have profound influence on the nature of growth-producing investment.

Figure 5.2 focuses on six power-related technologies. All are of proven utility in *niche* markets, but are still in stage-one innovation in the sense that each is considered to have a much larger potential applicability than at present, and all still require substantial cost-reducing breakthroughs in order to realize their full potential. Also, each presently requires slate of government subsidies to exist. Each cell of Figure 5.2 briefly expresses the implication of a dramatic decrease in the costs of applying the technology in the row in combination with the technology in the column. The diagonal cells correspond to each of the six technologies by themselves and the implication of a sharp lowering of costs associated with each of these but not in combination with any other of the six.

The most interesting thing about Figure 5.2 emerges from conjectures regarding which pair of technologies might emerge from stage-one. As a quick background, please recall that since the late nineteenth century, there has been a trend toward geographic concentration of electric power production. Indeed, the most important reason why power production is not even more concentrated in central power stations today is the limitations imposed by line losses of energy in the power distribution system. In the United States, 25 percent of power generated is lost as heat during transmission; hence the research and development interest in low- or no-resistance power distribution. Should this technology be the beneficiary of a cost breakthrough, there would be less need for power production facilities. Moreover, low-resistance transmission capability would mean that green power sources could be located in the most favorable situations (e.g., Southwestern deserts in the case of solar power) and the power generated could be moved to existing markets efficiently. A major breakthrough that reduces or eliminates line loss would encourage further *concentration* of power production.

Now consider another combination. The development of a small-scale power source that is economically competitive with today's central power stations could be combined with, say, greatly improved storage batteries to enable economical *decentralized* power production. Small-scale power could appear in the form of substantially improved solar collectors, fuel cells, or small-scale nuclear power plants.[20] The shaded cells of Figure 5.2 indicate combinations that

Figure 5.2: Enabling Potential of Future Dramatic Cost Breakthroughs

Cost breakthrough area

Cost breakthrough area	Local power production	Low-resistance transmission	Base-load storage-intraday	Solar collector	Wind power	Batteries, small scale
Local power production	Small Area Power, Substitutes for long-distance transmission	Not closely related	Enables economic small-area power production and consumption	Improved efficiency supports small-area power production	Economic power in small locations, rural?	Economic power in small areas
Low-resistance transmission	Not closely related	Substitutes for power production capacity by reducing line loss	Combination promotes intermittent power production: solar and/or wind	Enables location of intermittent power sources in favorable locations	Enables location of power source in most favorable locations	Not closely related
Base-load storage, intraday	Enables economic small-area power production and consumption	Combination promotes intermittent power production: solar and/or wind	Enables base-load solar or wind power	Combination makes solar a viable base-load power source	Storage during wind calms makes wind a viable base-load power source	Combination enables small-scale power production
Solar Collector	Improved efficiency supports small-area power production	Enables location of intermittent power sources in favorable locations	Combination makes solar a viable base-load power source	Efficiency improvement, does not affect intraday storage problem	Alternatives, not closely related	Comb. Enables small-scale power production: rural?
Wind Power	Economic power in small locations, rural?	Enables location of power source in most favorable locations	Storage during wind calms makes wind a viable base-load power source	Alternatives, not closely related	Efficiency improvement, does not affect storage problem during calms	Comb. Enables small-scale power production: rural?
Batteries, small scale	Economic power in small areas	Not closely related	Not closely related	Combination enables small-scale power production	Comb. Enables small-scale power production: rural?	Positive for electric autos, intraday storage for small area power production

Note: Shaded cells represent combinations that encourage reversal of the long-term trend toward centralized power production.

are consistent with a decentralized future. It is difficult to conceive of any development that would be more revolutionary than this, for it would enable a reversal of a trend that has been in progress since before 1900.

Thus, different combinations of ongoing stage-one threads can result in radically different futures. Needless to say, there are a number of combinations in Figure 5.2 that hold the promise of a technology revolution. There is not any really good way to impute probabilities to the possible combinations.[21] As a simple exercise, however, suppose that each cell has equal probability for realization in, say, the next five years. There are thirty-six cells, and of these, fourteen are conducive to a distributed power future. That leaves seventeen outcomes consistent with continuation of the trend toward centralized power production (the remaining cells are for combinations that add nothing to the effects of improvement of single areas). A number of pairwise combinations in Figure 5.1 have technology revolutionary potential, but of these, the combinations represented by the shaded cells may have truly huge revolutionary potential.

Why, one might ask, stop at pairwise combinations of dramatic cost decreases? Apart from the problem of presentation of a three-dimensional concept on a flat page, there is no reason, for combinations involving multiple technologies are conceivable, and the greater the number of these, possibly the greater will be the impulse to economic growth. Under the simple assumption of equal probability for each cell, Figure 5.2 portrays a game of rolling two dice. Admitting the possibility of three technologies undergoing dramatic cost reductions is equivalent to a game of three dice. This could happen, but under an equal probability assumption, the probability for each cell becomes 1/216 as compared with 1/36 in a two-die game. Again, the reader should be aware that equal probability is but an assumption and may have no justification in the realities of how technologies develop.

The enormous difficulties of predicting the main stream of technology future have not discouraged attempts to do so. To date, all such attempts have in common the assumption of the present state of technology. In recent times, some prognosticators have attempted to extend present technology by assuming near-term perfection of some or another existing technology that is still in its stage-one innovation. For example, one proposal for achieving a large supply of electric power from a totally renewable source calls for producing all power in a large photoelectric solar farm in the desert country in Arizona and New Mexico. This scheme assumes early perfection of a low-cost superconductor by which the power might be delivered to distant more densely populated areas of the country, as well as an economical means of intraday power storage. A variant on this theme dispenses with the dependence on further technological development. One variation, proposes direct solar heating of molten salt. The heat in the molten salt would be used to heat water in a heat exchanger which would be used to produce power conventionally, and for intraday heat storage. It would use high-voltage direct current (hvdc) trunk lines for low-loss power distribution to distant markets. Even though this variant requires no major future technology developments, it was estimated to be capable of completion only by 2050.[22] This type of scheme would be sold as an ultimate solution to green power, but the maturing of other technology combinations as in Figure 5.1 could quickly turn it into a huge white elephant.

It is unfortunate, but to date, the ideas of utopian visionaries do not get us very far in practice. It is tempting to suppose that this kind of futurist exercise is a by-product of present government efforts to force technology developments in directions that support policy objectives such as energy independence or greenhouse gas reduction. However, the utopian visionaries of today have plenty of company through history in failing to allow for the impact of technological change, for Karl Marx and T. R. Malthus based their futures largely on the state of technology that they could see.

The Private Sector as Fomenter of Revival

The Private Sector as Producer

Inasmuch as private sector investment is the preeminent force driving the upswing of the technology cycle, one might expect it to be an important element in initiating such a recovery. It can be, but only if it has some powerful new technology with which to work. Absent this condition, the private sector is crippled, especially in the opening stages of a deep recession following the maturation of the preceding upswing. There are three broad reasons for this. First, if the early stages of the recession are accompanied by malfunctioning financial markets, then the financing of any incipient investment surge can be seriously hampered. The negative impact of a dysfunctional financial market, especially the banking sector, disproportionately falls on smaller companies. Given the prominent role of new companies in past technologically based investment surges, this effect of malfunctioning financial markets is egregious. Second, there may not be a wide new field of technology immediately available for exploitation.

The third, and perhaps most important reason why the private sector is not a strong fomenter of early-stage recovery from a technology recession is that in the previous investment boom, established private companies tended to focus the R&D efforts on improving existing product lines. This activity is far less risky than gambling on the ultimate success of financing the stage-one innovation of a new and interesting technology. In 2010, a handful of large companies are avidly pursuing normal technological development of products which found wide markets in the 1990s boom.

Managements of companies that are established players in markets have a tendency to view radical new technology with extreme caution and sometimes as a threat to existing market positions. This is often true even when the new technology arises from the company's own research and development work. If this seems surprising, there is a reason for it. It often happens that established companies at the outset of a new technology revolution are the very companies that had their beginnings as small companies that spearheaded the *last* technology revolution. They have become completely adapted to the world of normal technology that grew out of the last revolution, and have become unfitted, managerially and organizationally, to benefit from an oncoming revolution, which appears threatening to the now mature company. The bankruptcy of Polaroid serves as a worst-case example of what can happen to an established player. The company aggressively spent on research, but this effort was concentrated on improving its main product, instant photography based on a chemical process. The development of low-cost high-quality digital photography negated anything that this company's research had accomplished.

Failure to have developed new product reflects a flaw of corporate governance: namely that the incentive structure typically faced by senior corporate managements discourages the kind of long-term thinking that would propel ongoing development of original invention and stage-one innovation and encourages thinking solely in terms of improving existing product lines.[23] Senior management compensation plans are at the heart of the problem. In general, these plans are designed by specialized consultants hired by compensation committees of corporate boards. The underlying goal of the plans is to create an incentive structure for chief executive officers (who are company employees) that will align their motivations with those of shareholders (owners). Which motivations? Inasmuch as the typical compensation package includes awards of company shares to senior executives, it is supposed that any action that promotes the value of these shares serves the interests of the larger body of shareholders. However, the day-to-day promotion of some metric deemed related to share value on a quarter-to-quarter basis, such as earnings per share, can conflict with consideration of the long-term interests of the company in an atmosphere of changing technology. Near-term considerations tend to triumph.

Important inventions with revolutionary potential have originated in the R&D shops of major corporations, but history is fraught with examples of managerial failures to derive full benefit from developments emanating from internal sources. Bell Laboratories effectively originated the transistor, but corporate management's vision failed to see the possibilities in this invention. The result was that the company sold the rights to the transistor cheaply to some smaller companies, such as Texas Instruments, that had a much larger vision as to future possibilities than did AT&T. Historically, the beginnings of technology-based recoveries have involved the efforts of new business organizations. Startups lack the large stake in existing markets that inhibits bold innovation by some established companies. The role of startups highlights the importance of maintaining legal and taxation conditions favorable to new businesses.

These patterns underscore the need for government policies that unambiguously encourage business start-ups and pose no barriers to the financing of these new companies. Such policies include moderate taxation, and, especially, the absence of conflict with government efforts at industrial policy that may go in directions other than the ones emerging. Government commitment to this or that gamble on technology developments are nothing but that: gambles, and even if a formal industrial policy is in place, its usefulness is closely related to its ability to recognize when it has made a losing bet.

Many of the fruits of emerging technology are delivered to the general economy by startup companies that become successful. These companies need new technology with which to work and face the difficulty of financing their operations while carrying out stage-one innovation on inventions. Their financing problems are greatest in times following a maturation of an earlier technology boom that ends in the midst of financial market collapse. Once the crisis is under way, the problem is not that of getting private managements to recognize the need for new products based on new technology, but to enable them to act by restoring the availability of credit on reasonable terms.

The Private Sector as Consumer

One of the early-on consequences of a financial crisis, such as that of 2007–2009 is an increase in consumer savings rates, or the equivalent, consumers' diverting consumption expenditures to paying down debt. The reduction in consumption is brought on by a number of factors, including loss of income due to loss of employment, the fear of job loss, and expansion of underemployment. These effects are not causes of the downturn, but are symptomatic of it. Impaired consumer willingness to spend is a threat to even a limited recovery from the recession; it is also a threat to business' willingness to undertake stage-two innovation, assuming that there are goods or business combinations available for such innovation.

Inasmuch as many technology upswings depend on expanding markets for the products and combinations which they produce, savings by unconfident consumers could be a problem in the initial stages of recovery. However, ongoing stage-one innovation of a technology or group of technologies is accompanied by falling costs associated with that technology. Indeed, falling costs are probably involved in the development of any technology-based upswing. Falling costs can be the basis for the start of a new consumer market even when consumers as a group are showing decreased willingness to spend.

Notes

1. The tax cuts of 2001 and 2003 were time-limited (set to expire after 2010) because of a quirk in the methods of projecting future government deficits. It was widely assumed that the cuts would be made permanent by a future Congress. As this did not happen, the cuts tended to lose any stimulative impact that they had when first enacted.
2. In the years immediately following the 2001 measures, their programmed expiration after 2010 was probably not a major obstruction to the perceptions of these measures as growth-producing. The programmed expiration was often dismissed as having resulted from an arcane feature of federal budgeting relating to calculation of future deficits. As time passed, however, Congress' continued failure to extend the 2001 rates beyond 2010 eroded such growth-inducing force the measures initially had.
3. Nathan Edmonson, *Technological Foundations of Cyclical Economic Growth* (New Brunswick, NJ: Transaction Publishers, 2009), 117ff.
4. Consider what the government's expenditure options are. Infrastructure is a leading example. However, in a country that already has a very highly

developed infrastructure, such as the United States, it is difficult to find a situation where an infrastructure investment creates productive wealth much in excess of what preexisted. What about proposals to spend for public mass transit systems? The United States once had much better mass transit facilities than it presently has, but most of these were scrapped by 1960 in favor of the private automobile. Massive growth since then has been in the context of the automobile, and has resulted in what is known now as "urban sprawl." Modern mass transit proposals are really a pathetic attempt to turn the calendar back because the payoff would require some very fundamental changes in population distribution. The time for mass transit investment to pay off can be reckoned, not in years, but in generations. Education is another object of government "investment." This amounts to throwing money at a problem in the absence of knowledge of the causes of the problem. This shows up readily in interstate comparisons, in which there is a large variability in expenditures per public school pupil that has no correlation with educational results according to standardized test results.

5.   This value reflects actual consumer reaction of the stimulus of 2008, which was in the form of a one-time income tax rebate; most of this was used to pay down debt rather than consume.

6.   Robert J. Barro, "Demand Side Voodoo Economics," *The Economists Voice'* 6, no. 2 (2009): Article 5.

7.   This proportion was officially 41.9 percent in November 2010.

8.   Milton Friedman and Anna J. Schwartz, *A Monetary History of the United States* (Princeton, NJ: Princeton University Press, 1963), 253–57.

9.   Access even to this kind of routine financing proved a problem in 2009.

10.   Mark-to-market accounting originated in the 1980s in the context of the trading of commodity futures contracts. The use of mark-to-market was seen as contributing to securities markets transparency, and spread to other markets such as markets for listed equities. Serious problems arose when the mark-to-market rules were applied to securities that are not actively traded (for which no market valuations are readily available or for which markets have become nonfunctional). Bankruptcy law contains many provisions by which many of the affected institutions could be reorganized successfully.

11.   This kind of argument from William Isaac and others was effective with Congress. As a result of the pressure of legislation, the Financial Accounting Standards Board (FASB) introduced changes in the mark-to-market rule that satisfied what members of Congress wanted. The proposed rule change would allow the holders of assets to use "internal models" to set values for assets held on their books when the markets were deemed to be dysfunctional and there is no evidence to the contrary. This rule proposal was dated April 2, 2009.

This rule change proposal has stirred controversy. Opponents see it as a total watering down of a rule that was originally designed to increase transparency in securities markets. Opponents' logic includes the idea that if mark-to-market was acceptable to financial institutions when the value of the assets in question were rising, then they should accept the consequences of a falling market. This logic fails to note that mark-to-market distorts the picture of institutional health in a down market, and therefore fails in its intent of yielding an accurate picture of condition. Questioning of the rule appears entirely appropriate on this ground.

12. William M. Isaac, Testimony before the House Subcommittee on Capital Markets, Insurance, and Government-sponsored Enterprises, March 12, 2009.

13. See note 11 above.

14. See note 12 above.

15. Eased mark-to-market requirements for distressed debt-related paper may work against as well as for the success of the Treasury's proposal to enlist private sector investors in relieving the banks of this paper. If a bank with substantial quantities of distressed paper on its balance sheet had the option of valuing such paper as the present value of an expected income stream, the result could be a higher valuation that is less threatening to the bank's solvency and therefore relieves the bank of some pressure to get such assets off its books. It is anyone's guess what this change would do to the process of negotiation between the bank and a would-be outside investor.

16. In both the GM and Chrysler cases, the United Auto Workers Union has emerged as a significant stockholder. There is an open question as to how this will affect the union's aggressiveness in opposing work rules changes that would improve efficiency.

17. This kind of governmental interference can already be seen in the cases of General Motors and Chrysler, especially: the government is heavily promoting cars that fit with its environmental policies, such as electric vehicles. This promotion appears to have little or no relation with what the car-buying public wants.

18. Interestingly, this plant is now operating profitably. When the private sector investors defaulted on their loans, the government sought to unload the plant which then appeared as a great white elephant. The plant was acquired in 1945 by a local utility for about four cents on the invested dollar, and is able to operate profitably with this new capital structure. Unfortunately, this "success" is not repeatable in the absence of some group stupid enough or weak enough (e.g., taxpayers) to absorb most of the initial investment.

19. E. C. Pasour Jr. and Randal R. Rucker, *Plowshares and Pork Barrels: The Political Economy of Agriculture* (Oakland, CA: The Independent Institute, 2005), 136–40.

20. When greenhouse gas emission began to be perceived as a problem, it was thought in some circles that this would bring a renaissance in nuclear power development, the appeal being the total absence of emitted pollutants with nuclear power. This renaissance has failed to materialize, however, because of cost: the cost of power from a traditional scale nuclear plant has been estimated to be much in excess of that from an advanced combined-cycle plant based on natural gas (the lowest-cost alternative) by the U.S. Energy Information Agency. Several designs for considerably smaller-scale nuclear plants (one-third of the full-scale plant) have been proposed. These offer several capital cost advantages. For example, they can be manufactured off-site and involve much less piping than does the large-scale plant, which has to be constructed on site. See Mathew L. Wald, "Giant Holes in the Ground," *Technology Review* 113, no. 6 (December 2010): 60–65.

21. One currently popular "scientific" technique for dealing with multiple futures is to express all the technology possibilities in a decision tree, and to assign (subjectively) probabilities to each branch of the tree. The result can be expressed in a histogram, whose mean becomes the single-number conclusion of the analysis. Inasmuch as the imputation of probabilities is educated guess at best, the technique offers little insight as to what will happen. The technique is widely used by security analysts. See, for example, Kenneth A. Posner, *Stalking the Black Swan* (New York: Columbia University Press, 2010).

22. Ken Zweibel, James Mason, and Vasilis Ethenakis, "A Solar Grand Plan," *Scientific American* 294 (January 2008): 64–73. The author has a simple system for evaluating such time estimates. If the claim is five years or fewer, then the presenter may have solid grounds for his/her estimate. If the number is higher than that, then it has little basis in science and is just as likely to be one hundred years or more.

23. There are exceptions to this general pattern. DuPont developed nylon and brought it to market during the great depression, for example.

6

The Technology-based Cycle: Obsolescence, Unemployment, and Recycling

Every piece of business strategy acquires its true significance only against the background of that process and within the situation created by it. It must be seen in its role in the perennial gale of creative destruction; it cannot be understood irrespective of it or, in fact, on the hypothesis that there is a perennial lull....
—Joseph A. Schumpeter

The U.S. economy in 2010 is experiencing a number of problems that are part and parcel of the transition from a period of revolutionary technological change to one of normal technological change. A technology revolution implies a surge of obsolescence, both of human and inanimate capital. It leaves in its wake a changed pattern of demand for materials. The theoretical economics literature has not dealt with the question of capital retirement under conditions other than those of normal technological change. In a technological revolution, the relation between capital retirement and the two main causes of capital retirement, wear and tear in use and obsolescence, becomes complex. Received theory also assumes that workers who become unemployed during recessions can be re-employed in a subsequent recovery that is induced by government countercyclical interventions; but these interventions make no allowance for structural changes in labor markets that result from technology revolutions. An ensuing technology revolution can bring a growth surge that will absorb unemployed resources, but it may occur only after a lengthy interval of normal technological development.

There is generally no significant restriction on a technologically based upswing due to resource scarcity. This, however, is not the same as saying that resources, especially labor, made redundant in a technology revolution, will be quickly re-employed during the reversion to a condition of normal technological development that follows the revolution. A changed labor market aggravates the unemployment problem, as discussed earlier in Chapter 5. A second problem is that a changed production paradigm can create changes in the pattern of demand for raw materials. This can include increased demand for raw materials not industrially emphasized in the technology paradigm that existed prior to the revolution. Adjustments in the supply of materials to accommodate a surge in demand in a changed production pattern include temporary shortages, price spikes, intensified searches for natural deposits, and recycling.

Recycling is not new, for there have always been active markets for reuse of common industrial materials such as metals and paper. What is new in 2010 is there exists an economic incentive to recover materials found in complex manufactured products, such as recovery of precious metals and rare earths from junked computers and other electronic products. In the past, recovery from complex manufactured products has been regarded as uneconomic.

Relation of Capital Retirement to Cash Flow

In light of often substantial resistance to adopting radically new technology by established companies, due to perceived threat of the new technology to established sales lines or simple underestimation of the potential of the new technology, the field of development for new technologies often falls to new companies. The appearance of powerful novel technologies creates favorable conditions for substantial change in industry structure. Indeed, as a general rule, the pioneer automobile companies were not prior builders of farm wagons.[1] Assuming these can find financing, startup companies have one advantage over established companies in not having a market position that could be compromised by success with the new technology.

In view of the foregoing, the question of how much capital is retired due to obsolescence in the year in which obsolescence becomes recognized remains. The data necessary for a straightforward statistical answer to this question for the most part do not exist. However, a logical argument sheds light on the question. Consider how capital

Figure 6.1
The Economy's Cash Flow and Capital Retirement due to Obsolescence

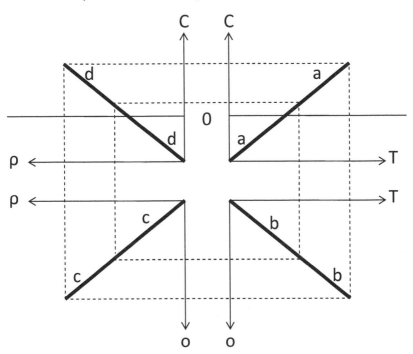

retirement as a percentage of current investment, designated by the symbol ρ, relates to cash flow, or C (Figure 6.1). A tentative conclusion is that the greater is C, the more capital-using firms are willing to retire obsolescent capital plant. The logic is that the higher is C, the costs associated with retirement of capital will have proportionately less impact on the bottom line, and the greater will be the ability to finance the purchase of replacement capital. To reinforce this logic, consider a kindred relationship: that between cash flow and the interest in using the latest technology, T. The upper right panel of Figure 6.1 depicts a positive relationship between C and T.[2] The lower right panel depicts the positive relation between the introduction of new technology and recognition of obsolescence, o. The logic is that if the technology available represents relatively small improvement over what is currently employed, the incumbent technology may or may not be replaced. However, if the new technology promises a spectacular improvement in operating costs, employment of it will create substantial obsolescence. The lower left panel depicts the relation between o and ρ. These three panels imply the upper left panel, which shows the positively sloped relation between C and ρ.

Retiring Obsolete Capital: Normal and Revolutionary Conditions Compared

As usually presented in formal economics literature, capital wastage occurs in the context of normal technological change. It refers to a decline in the economic value of a capital asset due to wear and tear during the course of the designed use of the asset. The asset is deemed to have an economic life, measured not in calendar time, but in units of operating time such as hours of operation or miles. The end of the economic life of a capital asset typically is recognized when its productivity has diminished to the point that the costs of restoration of productivity are sufficiently large that economic analysis indicates that the firm is better off putting the money toward a new machine. The incumbent machine is sold for its salvage value (scrapped) upon its retirement from productive use; it is withdrawn from the stock of productive capital.

Capital equipment becomes *obsolete* when a technically superior substitute becomes available. The concept of obsolescence includes the case wherein a machine of similar basic technology and function but with superior operating cost characteristics becomes available; it also includes the technological revolutionary situation in which new equipment of increased efficiency is available to replace not only an incumbent machine, but to replace the entire *productive system* of which the incumbent equipment is a part. There are cases in which emerging new technology becomes the basis for the formation of new firms in entirely new industries that compete with and possibly displace one or more established companies. When this happens, most or all the capital in the affected established industries becomes obsolete.

Normal Technology Conditions

In the simple case of replacement of a unit of capital with identical or incrementally improved equipment for doing the same thing, the concepts of capital wastage through use and capital obsolescence can appear to be similar. The research on capital replacement theory of the 1950s and earlier, treated the rate of capital replacement as a constant proportion of the stock of capital. In this body of theory, technological improvements in capital equipment could be expressed as an addition to the depreciation rate. Thus if capital wastage in year t in a technologically normal situation can be expressed as δK, where K_t is capital stock and δ is a constant, and if the steady rate of improvement in the productivity of new capital equipment is σ, then the combined effect of wastage and obsolescence could be expressed as $(\delta + \sigma)K_t$. The originators of this hypothesis made life easy on themselves by assuming that technological improvement in capital equipment proceeds at a steady, constant rate. The inadequacy of these assumptions[3] from an economic point of view was recognized during the 1970s.[4] At least some of these writers recognized that this treatment would not depict reality in the event of a very large increment of technology improvement, especially one whose effects include the obsolescence of the entire industrial system of which the equipment being replaced is part, and sometimes of entire industries in which it is used.[5]

The idea of obsolescence as a simple addition to capital wastage that could slightly shorten the economic life of a capital asset can be carried only so far. It comes closest to being an appropriate treatment in a world of normal technological change, in which all producers have time to adopt the most advantageous technology. In this world, productive capital is kept in service for all or most of its physically useful life, the rewards to replacement of capital plant early because of obsolescence being typically small. The principal reason is that the economic justification for replacing capital with technically superior capital before the end of its useful life typically rests on the belief that the new equipment offers the likelihood of increased profit. This means that the operating cost advantage to replacement has to be large enough to offset the increased depreciation charges that the new plant implies, and then some. In practice, this condition often proves to be a high hurdle for the possible replacement capital to overcome before incumbent capital becomes fully depreciated.

Technology Revolution Conditions

Consider the event that a number of radical new technologies that offer large cost or other advantages to adoption appear in a comparatively short time span. This might mean not just the replacement of a machine with another machine, even if the replacement is of substantially superior productivity, but replacement of the entire industrial system of which the replaced capital is a part. This is the case of the technologically revolutionary world.

The impact of personal computing on standard office operations provides a recent example of how a technological advance initially offered an opportunity to replace one element of a working system and then created the possibility for replacing the entire system. Prior to about the mid-1970s, authors of interoffice memoranda had to draft the memo with pen and paper and take the rough draft to a typist. Then the author had to revise and polish the typed draft, which the typist would then put into finished form for sending. In the late 1970s, a centralized printing machine became available. In this system, the typist would type the hand draft onto a CRT screen display and create a file. This speeded up the job of incorporating the author's revisions, as this could be done on the screen. When the author considered the memo to be final, the file was carried to the centralized printer, a large machine which produced a quality finished document. This change speeded the process from hand draft to finished document and rendered obsolete some conventional typewriters, but it still required a lot of human handling. Within several years, however, viable word processors for PCs appeared. This enabled memo authors to rough out and polish documents, and then print them. This change eliminated most of the time consumed in producing these documents. It rendered obsolete the central printers and many of the typists.

A classic example of this situation occurred in the 1900–1920 period, wherein a number of new energy conversion technologies became available. At first, the opportunity appeared to be for the replacement of stationary steam engines as prime movers of line-shaft factories; but as improvements in the power density of the new engines improved, the opportunity expanded into replacement of the entire line-shaft technology. What started as a simple equipment change-out ballooned into obsolescence of the entire line-shaft architecture, including the buildings which housed line-shafts and the supporting infrastructure.

While there is a well-developed theory of capital replacement in a technologically normal world, there is no corresponding body of theory of investing in a technologically revolutionary situation. The reasons for the neglect of the more dynamic case may have much to do with the lack of predictability of many of its elements. Concentration on the static case is often justified as a "normal" condition of replacement investment.[6] This is unfortunate. The occasional cataclysmic introduction of radical new technology such as that which enabled the IT revolution of the last two decades of the twentieth century, profoundly sets the nature and direction of the ensuing world of technology for decades. Moreover, the massive and prolonged capital investment necessary to implement radical new technology plus its required infrastructure underpins dramatic and lengthy growth episodes for the entire economy. Once the surge of investment opportunities stemming from radical new technology has become fully exploited, the general economy becomes susceptible to problems such as inflation, serious recession, and, possibly, depression.

Recycling of Productive Factors

One of the most important features of the long-term technology-based upswing is recycling of productive factors and materials in such a way as not to limit development of the upswing. Unfortunately, this is not the same as saying that all workers who have been displaced for reasons related to a technology revolution will quickly be re-employed. The key point here is that a major technology-based upswing lasts long enough that full exploitation of the possibilities that the technology revolution has created can be realized. The downside is that revolutionary change in a production paradigm profoundly affects capital and labor requirements in relation to what preceded the technology revolution. The retirement of obsolete or redundant productive factors during the upswing creates several problems. These include disposal of the physical capital that has been retired, and, more importantly, adapting of the workforces rendered surplus by the retirement of the facilities that employed them.

Plant and Equipment

Investment booms based on technology revolutions typically last for enough years that something approximating the entire range of profitable investment possibilities resulting from a technology revolution can be pursued. The usual concept is that obsolescent productive capacity will sooner or later be simply abandoned is partly accurate. Bethlehem, PA, was for many decades a steel-producing center, but the mill today sits idle. Its demise was the result both of its being dependent on the high delivered cost of raw materials and obsolete production facilities. This kind of result of obsolescence is more complex, however, than it might seem at first glance. A decision to abandon may evolve over a period of a number of years, just as a technology-based investment boom may develop its full force only according to a learning curve.

Inasmuch as the capital plant is often retired due to obsolescence following a technology revolution, it would seem as though opportunities for retaining retired plant in active production in a simple and direct manner are limited, but they nevertheless exist. There are many instances wherein capital plant is obsolete in the sense that there is an alternative production system available that would substantially reduce production costs, but whose economic justification rests on an assumption of continuous operation. In this instance, the newer system would be attractive as base-load capacity, but if the older equipment still works, it may still be useful as peaking capacity, especially if it has received proper maintenance. Petroleum refineries, for example, often kept their thermal cracking capacity operable as peaking capacity long after the adoption of lower-cost catalytic cracking capacity that became the base-load capacity. A number of railroads kept steam locomotives for overload and other intermittent service even after the base-load duties were taken over by diesel power.

Another circumstance that can keep older plant in productive use is that the overall economics of the productive system of which the plant in question is a part are dominated by some unique cost advantage in another part of the production system, such as superior access to some input. Natural resource industries' costs are often dominated by costs of basic inputs. One of the basic high-volume chemical building blocks is ethylene, for which the preferred input is ethane.[7] Ethane is produced with natural gas and separated from the main component, methane, for sale in the chemical market. Prior to about 2000, it was widely supposed in government and other circles that U.S. production of natural gas was following a long-term downward trend. This was taken to mean that the U.S. ethylene production, in order to remain competitive in a world market, devoted much research and capital investment into shifting input reliance away from ethane, and rendering the older ethane-based plants "obsolete." In those years, the supposition that ethylene producers in the Persian Gulf region had a strong cost advantage due to their access to low-cost ethane, and that future growth in world ethylene capacity would take place there. Since about 2005, however, this perception has begun to change radically. This change came from the realization that technological developments in the natural gas-producing industry have made vast quantities of gas deposits in U.S. shale formations economically accessible, and this implies vast supplies of U.S. sources of ethane. With this change in the resource market, the older ethane-based ethylene capacity has begun to look very valuable. Not only does the United States appear to be methane self-sufficient for the next century, but it also appears to be the low-cost source of ethylene. This change has huge implications for the location of future growth in the petrochemicals industry.

Several larger points emerge from the experience with natural gas production. First, what exactly is "obsolescence?" In this example, technology was applied to reducing costs in one aspect of a total production system—production of ethylene, and the need for "modernization" whose impetus was a perceived need to replace ethane as a feed is negated by the results of technological change in a completely different aspect of the same overall system—production of ethane.[8] Second, technological advance can result in some very fundamental changes in the economics of location in world industry. In this example, the implications are for some very basic changes in world trade in energy materials as well as in the locale of future growth in production of some widely used everyday products, such as polyethylene.

Scrappage of physical capital implies a write-down of whatever book value remains. The impact on profit and loss is negative, but the impact on cash flow is positive. This augments the cash flow available to finance investment in new-design productive plant. As pointed out earlier in this essay, a write-down is more likely to take place during the boom phase of the technology cycle because capital-using firms can most easily absorb the impact on the profit and loss account. Write-downs in deep recessions are less likely because of their impact on profits is proportionately greater than when revenues are cyclically high. Indeed, the economics of keeping incumbent capital in production in lieu of replacing it according to a fixed time schedule are probably more attractive in a deep recession than in a boom. Needless to say, retirements of productive capacity in a deep recession will commence with the oldest and least efficient plants.

Displaced Workers in a Recession

One overall characteristic of the upturn in the long technology cycle is that there are no serious capacity constraints that impede the expansion. However, this is not the same as saying that there are no serious unemployment problems in the immediate aftermath of a technology-founded investment boom, for there can be, as the

U.S. economy is presently experiencing. Broadly speaking, there are three reasons for the layoffs that occur in this circumstance. The first reason is straightforward: revenues have diminished, and fewer hands are needed to service the business that remains, even assuming no change in the organization of production. Reduced business directly affects production-line workers. The second reason is that the technologies that were adopted in the preceding investment boom enable reorganizations that allow the carrying on of a given level of business with fewer hands, such as by the elimination of one or more levels of middle management. The third reason is that the newly implemented technologies eliminate the need for some combinations of skills and experience while calling for different ones. The second and third of these reasons constitute a fundamental change in labor markets.

The rationale of government stimulus policies is open to question, but suppose for an instant that they work as their proponents suggest. In this event, workers can expect recall as product demand improves. Some of these workers often were not highly trained in the first place; indeed, the mass production industries hired such people because the work required only a minimal amount of training and experience. Retraining of line workers in mass production industries has not historically been a problem, provided the recessionary setbacks causing temporary diminutions of demand had little fundamental effect on the basic nature of production. Mass production industries have been able to draw labor from farms, immigrant groups, and other groups. The automobile industry of the 1950s and 1960s drew large numbers of coal miners who had been displaced by weakened demand in the coal industry. If the worker layoffs were due simply to reduced sales in a recession that otherwise leaves labor markets unchanged, stimulus measures might have a chance of working. In this mental experiment, a diminishing unemployment rate should accompany the more general recovery. So far, it has not. The problem is that the so-called economic stimulus measures that have been applied assume that *all* workers are like line production workers.

The unemployment that resulted from the recession of 2007–2009 was not a simple matter of temporarily diminished business level, however. Workers adversely affected by this business downturn also include highly trained professionals who worked in middle management and professional staff positions. Re-employing these can present a larger problem than re-employment of line workers. An anecdote illustrates this. In an annual meeting of the Society of Mining and Metallurgical

Engineers (SMME), this author attended a luncheon that was followed by a speaker, who was a retired president of a major mining company. When the speaker opened for questions, a geologist (suspected of being out-of-work) suggested that it might be wise for managements to retain geologists during slack times in order to have experienced exploration teams in place when business recovered (a labor-hoarding argument). The answer, while sugar-coated, came down to this: (1) geological teams could be reconstituted with recent graduates who would probably be cheaper and more up-to-date professionally than senior geologists and (2) the geology profession was overpopulated to the point that replacement was no problem.[9] Such an answer could equally apply to any number of engineering specialties, and to other professional groups. Unemployed people in this group are not exactly displaced technologically, for their skills are still needed; they are simply surplus.

The third affected group of unemployed workers is composed of people who had significant job-specific skills no longer needed in the context of the changes in labor demand wrought by the technology revolution. These people can be considered unemployed structurally. The important question is concerned with how large is this structural component in overall unemployment, for their re-employment in jobs similar to the jobs they held at the time of their dismissals is unlikely. Such jobs no longer exist in many instances. This question goes to the heart of the notion that ordinary countercyclical measures can bring reduction in the unemployment rate.

One key question in the aftermath of the recession of 2007–2009 is the extent to which unemployment is cyclical or structural. In the latter event, it is not likely that conventional monetary or fiscal policy tools will have much impact on unemployment. From the standpoint of this question, the second and third of these unemployed groups (those that the boom rendered redundant or obsolete) is of high interest. The clues to the importance of these groups are (1) the historically high proportion of people who have been unable to find jobs in twenty-seven weeks and (2) the high number who have dropped out of the labor force or are underemployed. The IT revolution was different from earlier technology revolutions in that it brought substantial productivity improvement in what are generally regarded as service industries. A recently published study has documented the IT impact on the service sector. From 1987 to 1995, overall service sector output per worker-hour increased at an average annual rate of 1.93 percent per annum; but from 1995 to 2001, this rate *accelerated* to 2.63 percent per annum. An examination of the industry detail included in these aggregate figures reveals that the productivity growth occurred in a wide spectrum of service sector industries.[10]

Prior to the IT revolution, it was common among economic analysts to think of productivity change as mostly confined to goods-producing industries. Therein lies a possible clue as to the importance of the structural component in the current level of unemployment. The improvement in service sector productivity was a natural consequence of the nature of the IT revolution, whose hardware, and software developments created new possibilities for ways of operating successfully with fewer hands. In many cases, these opportunities appeared for the first time. People surplussed in this process were let go. Their jobs are probably gone forever.

Historically, governments have attempted to provide formal retraining for surplussed workers, but these programs have often failed on the grounds of high costs and ineffectiveness.[11] There is a large retraining industry in the private sector that includes evening courses at colleges and technical schools. These can be costly, but can be attractive to younger workers displaced due to obsolescence or redundancy.[12] Older workers clearly have fewer options than younger ones. Often, the upgrading of the labor force becomes a generational process. A sad (but typical) example comes from the steel industry of the Mahoning Valley of northeast Ohio. In the past, it was a common occurrence for young men to go to work in a nearby mill upon graduation from high school, and for years it was not unusual to have two or three generations of a family employed in the local steel industry, sometimes in the same mill. Upon the postwar decline of the area steel industry, children of steelworkers began to seek wider opportunities out of necessity. Many moved to other areas of the country. Older workers experienced increasing distress. To continue the steel example, workers past fifty years of age who had no experience other than steel mill work faced huge problems in retooling their skills, including a high cost of retraining and limited time in which to benefit from retraining.

Recycled Materials

As noted in the introductory paragraphs to this chapter, there has always been recycling of various industrial raw materials such as copper, steel, and paper. The recovery from the 2007–2009 has seen an emphasis of materials recycling that is novel by historical standards.

Three elements enter into and cause this. First is price. In a functioning commodity market, an increase in quantity of a commodity demanded will be reflected in rising price, and the response to this is increased quantity supplied. Second is the changed industrial paradigm, which creates demand patterns for materials different from what preceded the technology revolution. Third is technology itself, one of whose effects is physical concentration of material supplies.

The third of these elements, concentration, may be the most important of the three in the explanation of what is happening in early 2010. Consider, as a mental experiment, the question of which is less costly: production of aluminum from its ore (bauxite) or from discarded aluminum beverage containers? The obvious-seeming answer is the latter, for it is already aluminum, and reclamation does not require the costly steps of mining, refining, and smelting that the former requires. However, this is the answer *only if sufficient tonnage of recycled beverage containers can be delivered to the smelter gate sufficiently cheaply.* Inasmuch as discarded beverage containers are found in small quantities in household refuse and along roadside ditches over the entire country, the cost of bringing the needed tonnage to the smelter from these widely dispersed sources would be huge, and if it had to be absorbed by aluminum producers, aluminum production economics would tip strongly in favor of the ore–alumina route to finished aluminum. As things are in 2010, the concentration of recyclable aluminum is carried out virtually for free by thousands of government park authorities, civic groups, and individual households, all of whom are responding to a desire for a pristine environment. The popular interest in a pristine landscape that this evidences is the historically novel element in today's recycling.

Concentration is the key to any decision to produce a material from scrap or from de novo sources such as ores. Concentration largely achieved through citizen efforts, as in recycling of aluminum, should be regarded as a special case; most recycling of any industrial importance has historically been accomplished by a for-profit recycling industry.[13] Concentration has been achieved in two ways. In the first, the economics of recycling a number of common industrial materials have been dominated by demand for the final products derived from the materials in question. No better example exists than copper and nonferrous metals, whose major markets have long been characterized by cyclic movements. When the price of some finished product of, say, copper, is high, the price of scrap also rises to the point that it is economical to collect and market it.[14] The scrap market for other recyclable materials has always been affected by similar cycles. In general, a price boom in a recyclable commodity will attract concentrated sources of material for industrial processing.

In the second basis for concentration, technology plays a crucial role. There are some interesting modern examples of this. The current auto industry interest in electric power for automobiles has resulted in a derived demand for permanent magnets for use in high-end electric motors, and this has resulted in a large increase in demand for certain rare earth metals, such as neodymium. This increase has inspired not only a massive exploration effort for natural deposits, but also a subindustry of the recycling industry to recover this metal from scrapped objects. Inasmuch as the material is a minor component of the scrapped goods, the new activity has taken the form of dissecting these scrapped assemblies in order to reach the desired materials. Increases in the prices of precious metals and rare earths have recently fostered an industry devoted to recovering these materials from composite products that have been junked.

Notes

1. Studebaker was an exception to this generality.
2. Note that cash flow is shown as having a negative range. For the economy, this condition could be indicative of deep recession conditions. It also allows for the case in which a group of startup firms having little or no revenues are engaged in stage-one innovation of new technology, a situation often found in the early genesis of a technology-founded investment boom.
3. This treatment assumes that technical improvement proceeds at a constant rate, and that capital is utilized at its design capacity over its entire economic life. However, actual operating lives can vary significantly from hypothetical lives measured in calendar time. For example, a period of low product demand can involve reduced utilization rates of productive capital, and this extends life measured in operating time units, such as hours or miles. See Martin S. Feldstein and David K. Foot, "The Other Half of Gross Investment: Replacement and Modernization Expenditures," *Review of Economics and Statistics* 53, no. 1 (February 1971): 49–58.
4. This treatment of the combined obsolescence and capital wastage assumes that wastage is a mechanical process entirely determined by physical characteristics of the capital itself, and that obsolescence reflects technological progress at a constant exponential rate. A number of writers saw this treatment as not only unjustified by observation, but actually misleading. See for example, Martin S. Feldstein and Michael Rothschild, "Towards an Economic Theory of Replacement Investment," *Econometrica* 42 (May 1974): 393–416.
5. See Vernon L. Smith, *Investment and Production* (Cambridge: Harvard University Press, 1966), 128f.

6. Ibid., 144–45.

7. A low-cost ethane source permits use of a relatively simple plant for producing ethylene. Alternate feeds require more complex, therefore more costly, plant.

8. The crucial element in anticipating highly competitive costs in a U.S.-based ethylene industry based on ethane is low delivered cost of the ethane feed. Much domestic ethane is produced near to a well-developed pipeline system dedicated to the movement of ethane. This results in low delivered costs of ethane even in comparison with competitive ethylene plants in the Persian Gulf region.

9. At least some of this condition can be laid at the door of the professional counseling that students received during their university experience. The chances are good that the counsel was provided by tenured faculty with a strong interest keeping students in the department's classes. This motivation may not always square with the student's interest.

10. Jack E. Triplett and Barry P. Bosworth, *Productivity in the U.S. Services Sector* (Washington, DC: Brookings Institute Press, 2004), 18–24.

11. For example, the infamous CETA program of the late 1970s.

12. Worker displacement can be a direct result of technology as in a declining industry. It can also be due to the proximate cause of offshore movement of jobs. In this case, technology may be the ultimate cause, as some U.S. industries are susceptible to competition from foreign factories using a combination of cheaper labor and advanced technology. It is true that advanced technology can be applied in the United States, but such application often means reduced labor force.

13. While there is a paper recycling industry, it has been aided by volunteer efforts, especially during paper price surges. Volunteers include Boy Scout troops and churches.

14. Copper price surges have induced criminal activity, in which individuals loot copper plumbing and wiring from unoccupied buildings. This is very much a problem today.

7

Conclusions: The U.S. Economy in the Early Twenty-First Century

Relevance of the Technology Cycle

In 2010, the U.S. economy, and to some extent, the economy of other industrial countries, are at the trailing edge of a capital investment boom that grew out of the IT revolution. Even though the recession was over in 2009 by the National Bureau of Economic Research measure, the recovery remains weak by historical standards. While corporate profits have rebounded, the unemployment rate remains only slightly below where it was at the bottom of the recession. Reignition of vigorous growth at levels experienced in the 1990s has been the objective of government stimulative efforts, but none of these have had much effect to date, unless these be credited with the anemic recovery that actually has transpired. Inasmuch as a collapse in housing values was the proximate cause in bringing on the recession in the first place, public authorities have pointed to housing market recovery as a necessary condition for general economic revival; however, the problems still besetting the housing market do not auger well for near-term recovery. Moreover, the roots of the present weakness are deeper than housing.

An economy of the complexity of that of the United States is a living organism and is constantly evolving. Its growth is part and parcel of this evolution. Economic growth has never occurred in a uniform manner. It has occurred in distinct pulses which are based on technological revolutions. These pulses can be visualized as temporary accelerations of ongoing evolution. Three such pulses can be identified during the entire twentieth century. A technology revolution, such as the IT revolution, presents a large slate of productive investment opportunities, but when this slate has become largely exploited, the economy is vulnerable to a setback. It also presents an overall production paradigm that is different from what preceded the technology revolution. The 2007 recession was triggered by the failure of a residential real estate bubble, but the weakness of the recovery is indicative of a temporary paucity of productive investment opportunities.

Between growth pulses, which is the situation of the U.S. economy in 2010, technology advances relatively slowly. Technological change under current conditions consists of incremental improvements of technologies that emerged from the IT revolution. The potential for improving any one stream of technology declines in the long run, and technological stagnation is the expected result if the period of incremental improvements lasts long enough. However, the technology revolution, which is based on some fundamental improvement of understanding of how nature works, plays the role of a resetting of the entire technology process by creating an entire new slate of improvement potentials.

If the economy of the United States—and the developed world—depends on sometime technology revolutions, what is the probability that these events will recur in the future? The overall evidence suggests that the pulsating character of past growth has not been merely the accidental occurrence of runs in a coin-tossing game, but has been a systematic feature of economic growth. Inasmuch as recurrences of technology revolutions of the past have occurred at intervals ranging from twenty to thirty years, it may be supposed that the next one will occur sometime according to this admittedly imprecise timing. It is true that there is plenty of activity that can be termed "technological" at the time of this writing, but so far, none of these technological threads has met a crucial precondition for the launch of a technology-based investment surge: namely, a cost breakthrough in a group of technological threads that are not at the center of the established pattern that offers the potential of radical change in the way people live.

Present and Ongoing Problems

The state of the long technology cycle profoundly affects the economy's reaction to applications of standard tools of government countercyclical policy. It comes simply to this: if there is an investment boom based on a technology revolution in progress, then monetary policy will be effective at countering negative developments that may arise from instabilities in financial markets; but if no technology-driven investment boom is in progress, then an expansionary monetary policy will not foment renewed growth and may render the economy highly prone to inflation and asset valuation bubbles. The economy's ability to absorb liquidity into productive assets varies over the technology cycle and is low immediately following a technologically based investment boom. Clearly, the long technology cycle deserves far more attention than it has heretofore received.

Persistent Unemployment

The stubbornly high rate of unemployment is a conspicuous reminder of the recession of 2007–2009. The causes of this persistence can be summarized quickly. A technology revolution, such as the IT revolution, profoundly affects the nature of labor markets, which are different from what they were prior to the preceding investment boom. Among other things, the IT revolution presented a range of opportunities for producing a given output with fewer hands and different arrays of skills, for this is the broad implication of the increase in worker productivity that a technology revolution creates. In a recession that immediately follows a completed technology revolution, a recovery that is of sufficient vigor to provide jobs for all those who lost jobs is unlikely. Only another technology revolution will accomplish that.

The majority of U.S. workers are employed in what is generally called the service sector. In the past, it had been habitual to think of technological progress as applying primarily in the goods-producing sector, but this has changed. The hardware and software that emerged from the IT revolution found efficiency-enhancing applications across a myriad of different economic activities that come under the general heading of "services." A further hint of the structural nature of present unemployment is an unusually high percentage—by historical standards—of unemployed workers who have been unemployed for more than twenty-six weeks. It has been remarked that lengthy episodes can impair a worker's ability to find productive employment in today's industry. At such time as private companies recognize a need to expand employment, there is an open question regarding how they will accomplish this: by hiring newly minted talent in the form of recent graduates, or by bringing back previously laid-off workers.

163

Corporate Profits

While a broad spectrum of corporate profits was negatively impacted by the recession, recovery in profitability has been stronger than for the economy in general. Sales recovery has reflected general economic conditions plus some exports, and profitability has been strongly helped by cost cutting, including worker layoffs. This is consistent with the idea that the IT revolution presented industries of all kinds—whether goods or services—with opportunities to maintain a given output with fewer hands. At least a part of the currently and recently favorable corporate earnings reports can be attributed to the multinational character of industrial business. While North American sales may be down, a number of companies are doing reasonably well with their operations in certain emerging foreign economies. However, it can be assumed that as long as general revenue growth weakness persists, established enterprises will have little incentive to expand U.S. employment on a sufficiently large scale to bring a significant reduction in the unemployment rate.

The weakness in private investment is a central point of concern, for is it not private investment that creates productive assets and therefore jobs? Government officials criticize private firms for holding large cash positions instead of investing in productive assets. But in what productive assets? In the 1990s, the economy benefitted from heavy capital investments based on the IT revolution. Investment demand originated not only in expanding demand for the products of the IT revolution, but also from demand for production facilities for manufacturing these products and for replacing the productive capital obsolesced by the IT revolution. Inasmuch as the revolutionary productive capital of the 1990s is now commonplace, present-day investment is necessary only to replace worn-out capital with what is, at best, only incrementally improved versions of what was acquired in the height of the investment boom. This level of investment is necessary, but it is not sufficient to support the level of investment that created the job growth of the 1990s. It is a sad reality that the recovery so far, such as it has been, is probably the best that can be achieved in the absence of a renewed support from a technology-based investment boom.

New Business Formations

In spite of favorable earnings reports from established companies, especially those with international operations, the private business sector still shows at least one glaring weakness: a dramatic drop-off in net new business formations. To see why this is a matter of serious concern, one only has to reflect on the role of new companies in developing and applying the tools of IT that underpinned the IT revolution of the 1990s. While a number of elements have contributed to this fall-off, such as recent legislation that imposes proportionately heavy cost burdens on smaller public companies, the single-most compelling probable reason for the fall-off is the absence of emerging new game-changing technology that is economic. This may seem a strange thing to say, for the technology industry appears alive and active. However, there is a tendency for recent newly organized companies to be responses to government subsidies, and whose prosperity depends on continuation of such subsidies. This is much different from new company formation on the basis of opportunities posed by new technology. Today's subsidized new starts include corn ethanol and battery enterprises. So far, there is no obvious cost breakthrough in any technology field that could trigger a growth pulse.

Historically, large companies have sought stability above all else, and this is the reason why such a large part of the onset of a technology revolution features new companies. This historical pattern has changed a little, in that a number of large companies have bet heavily on the passage of legislation of an environmental character, such as a carbon tax or equivalent. In the so far absence of such legislation, some of these bets now look less than inspired. In many cases, today's large and established companies are the very companies that had their start and initial expansion in the years in which the IT revolution itself was forming. As new companies, they were bold; as established companies, they are cautious and conservative.

Residential Housing and Commercial Real Estate

Failure of what now appears to have been a massive bubble in housing assets was a proximate trigger of the deep recession that followed 2007. As of 2010, the housing market has not shown much sign of recovery; indeed, housing values continue to decline in many area markets, as well as the value of financial instruments derived from the value of the housing stock. Sales of existing homes have declined sharply, and widely held beliefs that the bottom of the housing market is still ahead are not helping this market. It suffers from a large overhang of unsold existing homes as well as an overhang of homes whose owners desire to sell, but have been held off the market pending an improved market. Naturally, new home construction activity is well down.

When the housing market failed, holders of many extant mortgages and derivatives therefrom found their balance sheets afflicted with "assets" with little or no short-term value. Government proposals/ efforts to restore lender soundness by acquiring these "toxic" assets foundered over the issue of how such assets should be priced. The over-all result of these efforts is that many assets of questionable value are still on the books of financial institutions and investors. To date, there have been no measures that would relieve banks of these assets without threatening the solvency of a large share of the banking system. Indeed, the one event that definitely would right this situation is a revival of housing values, and that does not appear to be a near-term prospect. The latest proposal for dealing with this weakness is the so-called QE initiative in which the Fed buys up to $600 million in U.S. securities from banks. This program appears to be aimed at improving banks' capital position to the point that they consider themselves able to write down their inventories of toxic housing-related paper without threatening their own solvency. As noted below, however, QE risks promoting serious inflation. The Fed dismisses this danger by pointing to the high unemployment rate, saying that it is a cyclical phenomenon. They do not seem to consider that current unemployment has a large structural component. If high unemployment is structural, its capacity to cushion against inflation is much more limited than the Fed leadership supposes.

To appreciate fully why the depressed conditions in the housing market are such a threat to healthy recovery, one only has to do the mental experiment of supposing that the housing market returns to equilibrium. One very important result of such a revival would be an upturn of the housing construction industry. This would enhance the economy's ability to absorb liquidity productively. This is the reason why housing construction revival has led the economy out of a number of past recessions; it approximates the effect of a renewed technology-based investment boom, at least for a while. As things presently stand, it appears likely that, in general, housing values have to fall further in order to right this market.

In many ways, the commercial real estate market is in little better condition than that of residential real estate. Many commercial proper-ties whose market value has fallen sharply are carried on the books of their owners at historical costs rather than market value. This practice is permitted by an accounting rule which says that if an asset is held as a long-term investment and there is no intent to trade it in the near term, then there is no pressure to revalue it to current market value. This practice has so far masked the true condition of the market and has forestalled a collapse parallel to what happened in the residential real estate market.

Government Measures

Countercyclical Fiscal Policy

Apart from monetary policy, which could be an effective tool for maintaining stability in the intervals between technology-based investment booms, there is countercyclical fiscal policy, based on the idea that growth can be stimulated by government deficit spending. Such measures have not been deemed necessary to offset any but the most serious of recessionary setbacks, but setbacks of this magnitude tend to be those following technology-founded growth pulses, circumstances in which fiscal stimulus is least likely to have any positive growth effect. These measures retain adherents among some academic economists despite ample evidence from a number of countries as to lack of effectiveness. One is tempted to the idea that the evolution of economic theory has not kept pace with that of the economy itself.

The idea of reviving vigorous economic growth via deficit spending is especially alluring to politicians, whose first impulse in any crisis is to be perceived as doing something. Thus even though academic believers in the efficacy of fiscal stimulation may disappear one death at a time, politicians are too fond of the idea to give it up any time soon. The prime exhibit of the ineffectiveness of fiscal stimulus, as usually conceived, is its failure to bring much reduction in the unemployment rate that developed following the great recession, from which the recovery remains anemic.

This is not to say that government cannot make a meaningful contribution to economic stability from spending policies. The key is the continuous and ongoing support of technology research and development. It is easy to point to substantial elements of the IT revolution that were born as government research activities. Two that come readily to mind are the Internet and global positioning. There are numerous others as well. Interestingly, these two major examples came not from deliberate efforts to stimulate the economy, but from narrow Department of Defense objectives.

It is difficult to cite an example of a successful government economic stimulus program in any country that was initially conceived as such. The most effective government contributions to economic health tend to have come from spending for seemingly unrelated purposes, such as national defense. Among the least effective government stimulus programs are those that come under the heading of industrial policy. There are a number of reasons for this unhappy fact, but of these, a leading one has to be that these efforts often involve intensifying a research focus on some particular line of technology that is still in stage-one innovation. Current examples include the effort to develop the so-called renewable sources of electric power into base-load power sources. Technology never has responded to such spending, inasmuch as the cost decline that is critical to making a technology commercially successful does not respond predictably to having money thrown at it; nor is it likely to do so in the future. The research process grinds onward, and the best the government can do is not to obstruct its progress through ill-conceived policies of taxation, regulation, and industrial policy.

Tax Policy

A moderate rate of taxation can be a necessary (but not sufficient) precondition for the development of a technology-based investment boom, especially if the tax structure is easy on incomes derived from savings. Substantial tax cuts in 2002–2003 were timed to expire at the end of 2010. This timing had to do with the manner in which future deficits are estimated, and the expiration timing was solely for the purpose of making the projected deficits appear less formidable. At the time of enactment, tax cut proponents assumed that a future Congress would at least extend the cuts beyond 2010 if not make them permanent. At the time of this writing, they have been extended to 2012, ostensibly to avoid what would have been an abrupt and steep tax increase that threatened to upset the anemic recovery from the great recession. The extension grew out of a political compromise in which opponents of continuation supported extension very reluctantly.

Tax reduction can be a powerful stimulus to economic activity, but to realize this benefit, the cuts have to be perceived as permanent. The 2003 cuts may have contributed to the recovery from the recession of 2000–2001, inasmuch as there appeared to be a realistic hope of their becoming permanent, but by 2007, this hope was largely faded, and any simulative effect from the relatively favorable tax treatment of savings income was swamped by the collapse of the housing bubble. The two-year extension enacted in late 2010 does nothing to ease uncertainty regarding future rates of taxation, and the prospect of a sharp increase in the taxation of businesses and savers in the foreseeable future inhibits willingness to invest, even when productive investment opportunities can be found. Arguably, the uncertainties created by the temporary nature of the present U.S. tax system discourage investment.

Much of the argument for letting the expiration of the 2002–2003 tax cuts take place on schedule was based on the contention that the cuts favor incomes derived from investment. In other words, the cuts favor the "rich." Arguably the worst use of tax policy is as a tool of social engineering, which is the common justification of tax policy as a tool against income inequality. Even if this were a worthwhile goal, tax policy is a very clumsy a tool; and even if it were an effective tool for income equalizing, the purpose itself is of questionable value. Rapid growth in an economy as large and sophisticated as that of the United States, such as occurred in the 1990s, inevitably produces income inequality. Inequality diminishes, however, during periods of normal technology change. A direct attack on income inequality is a direct attack on a symptom. Physicians know that one cannot treat a melanoma with a band-aid, but this is what attacking income inequality amounts to.

The U.S. government has resorted to temporary measures such as investment tax credits and fast write-offs of capital investment expenditures, and these have appeared to be effective at stimulating business investment for short periods of time. The weakness of such measures is obvious—the effects are temporary. Their justification is usually of the "pump priming" variety—in which the investment induced will generate enough overall incomes via the multiplier effect that the tax incentives can be withdrawn after minimal loss to the government's revenues. This argument fails on the grounds that investment is of itself not a fomenter of growth; it is only the mechanism by which the benefits of advancing technology become transmitted to the wider economy. If business tax incentives of these kinds are imposed when there is a paucity of productive investment opportunities, such as following the completion of an investment boom born of a technology revolution, then their effects will be weak.

One aspect of tax policy that could be changed constructively is a lowering the marginal rate on corporate profits. The most compelling reasons for this suggestion is international competition, wherein a handful of competitor nations impose lower rates than the United States does. This is by far not the only factor influencing corporate location, for there has so far not been a rush offshore of major U.S. corporations; just a seepage. The threat here is long term, and this is unfortunate. It makes it too easy for revenue-seeking governments to put off dealing with the problem. It goes against populist arguments that derive from desires to reduce income inequality, and such populist arguments will always find adherents among politically ambitious individuals.

Then there is the problem of tax policy's influence on savings. As every beginning economics student learns, the collected savings of individuals and business support the financing of capital investment, and clearly play a crucial role in the financing of a growth-inducing technology-derived investment boom. However, textbook diagrams showing the relations between savings and investment, taken literally, apply to a closed economy. There are no important closed economies in the world today. In today's world, financing for a successful new company can come from a variety of sources outside the United States, and, indeed, given the low individual savings rate in the United States in recent decades, foreign financing sources have become very important. There is ample evidence of popular nervousness at the idea of selling U.S. assets into Chinese or Middle Eastern ownership.[1] However, this has not so far become reflected in U.S. tax policy, which in many ways discourages savings.

Savings are an important issue in 2010. Prior to 2007, the savings rate as calculated in the N.I.P.A. accounts was very low in historical terms. Consumers had fallen into the habit of saving little or none from current incomes, believing that increasing values of their residences took care of perceived savings needs. The collapse of housing values disabused many of the idea that their houses could be used as ATM machines, and emphasis shifted to restoration of household balance sheets. Personal consumption fell. Available income flows became dedicated to debt reduction. These events are the equivalent of a sharp increase in personal savings rates. Government fiscal policies aimed at restoration of consumer spending assumed that once consumers regained their feet financially, they would resume pre-2007 spending habits. This is a questionable assumption. It amounts to thinking that consumers will quickly shed the memories of the housing collapse. However, a trauma of this magnitude is unlikely to be quickly forgotten.

Monetary Policy

Monetary measures have come to be at the heart of government tools for offsetting the effects of recessionary setbacks. However, monetary countercyclical tools work best when the recession is a simple interruption in a technology-based cyclic upswing because labor markets after the recession are little changed from what they were before the recession. Monetary stimulation has *not* been effective in bringing a vigorous recovery from the downturn of 2007–2009 because one of the most important results of a technology revolution is that an ensuing increase in private sector unemployment is *structural* in nature; it is not simply cyclical. Therefore, *countercyclical* monetary stimulation is not going to have much effect on reducing unemployment. In a technology revolution, producers learn how to produce more with fewer hands. In a recession that follows a technology-based investment boom, they have little incentive to pad employment rolls, inasmuch as they have gained the ability to accommodate the relatively mild demand increases characteristic of normal technology change with reduced employee headcount.

The Federal Reserve's bemusement with the idea that expansive monetary policy can induce economic growth in and of itself partly arises in its charter, which directs it to pursue full employment and stable prices as policy goals *simultaneously*. When unemployment contains a substantial structural element, however, then countercyclical easy money policies are ineffective against unemployment but risk inflationary mischief. The two prescribed goals are contradictory. The simplest way to reduce this conflict would be for Congress to revise the Fed's charter so that it can focus solely on price stability. Inasmuch as vigorous growth resumption is unlikely to resume in the absence of a strong technology push, maintenance of price stability ranks with light taxation as a condition favorable to the development of a growth-inducing investment boom at such time as the technology basis for it emerges.

The effects on the general economy emanating from the technology cycle, as herein described, have never had much of an impact on the formulation of monetary policies. Inasmuch as the state of the technology cycle very much affects the effectiveness of monetary measures, this is unfortunate. There are possibly a number of reasons for this neglect, but of these, the most important has to be that the technology cycle operates as a basement under which financial and money markets operate, and is not always clearly visible; it is too easy to ignore, and customarily is. Present and recent monetary policy measures follow the customary historical pattern for monetary policy to be expansive well beyond the completion of technology-based investment booms in the hope of inducing investment at a volume sufficient to encourage productive investments. This policy is futile when the economy has temporarily depleted a stock of productive investment opportunities, but the policy does create conditions favorable to the formation of asset price bubbles, some of which, such as housing, exist at the very heart of the economy.

What has the Federal Reserve done so far? The main policy tool to date has been interest rate management, and short-term rates have been essentially reduced to zero. However, this is not the limit of the Fed's tool chest, or so the Fed leadership claims. A current policy, termed "quantitative easing (QE)" consists of massive Fed purchases of medium- and long-term government securities from member banks—and possibly newly issued Treasury debt. This is purely "printing money." Inasmuch as a prior condition of historically high liquidity in the economy has so far failed to stimulate a strong recovery, the QE proposal amounts to doubling down on what increasingly appears to be a bad bet. The sooner-or-later inevitable result will be a serious inflationary episode. The Fed's leadership has played down the inflationary potential of QE by pointing to the high level of what they are calling "cyclical" unemployment. Because the current unemployment situation has a large structural component, calling it cyclical sounds like a labeling of evidence to support a preordained conclusion.[2]

U.S. monetary policy presently faces a serious dilemma. To raise the interest rate structure now poses a threat of triggering a second round of the recession. Rate increases should have commenced well before 2005. Pursuit of QE, alternately, promises serious inflation. This is a lose–lose situation. Monetary tools today are being used to deal with the wreckage of financial markets after the collapse. The experience of the recovery so far highlights the futility of this alternative. A scenario of future combination of high unemployment and inflation is a real possibility.

Just what is a proper countercyclical role for monetary policy? Its overall objective should be to maintain economic stability in the economically vulnerable time periods between the completion of one technology cycle and the commencement of the next. Inasmuch as a technology revolution generally cannot be induced, this assignment amounts to the task of preserving a semblance of prosperity for an indefinite period of time with little help from technology progress. Historically, monetary policy following what can be recognized retrospectively as investment booms based on technology revolutions has tended to be expansive, characterized by low interest rates and a high degree of liquidity in the economy. This is equally true of the late 1920s, the 1970s, and the years following the recession of 2000–2001. The goal of expansionary monetary policies is to foment investment, but if slate of attractive opportunities in productive investment has become fully exploited, there is the question: investment in what? The problem after the investment boom of a technology revolution is a dearth of productive investment opportunities beyond the requirements of normal technology, *at best*. Even if revenues have not totally crashed in the financial setback, the remaining private investment opportunities are those of normal technological change, and these are not sufficient to foment a serious growth impulse.

Uncertainty

If the positive effects of government countercyclical measures to date have failed to induce renewed vigorous growth, it is natural to ask what other effects these measures might have had. The object of countercyclical policy has been to increase the liquidity in the economy on the assumption that the easy availability of money will induce consumer spending at pre-recession levels and private business investment to accommodate anticipated increased levels of business. This is not happening, and is unlikely to happen for several reasons. The recession and its aftermath have been traumatic for many consumers. The great depression produced a whole generation of frugal savers. Will today's debt-burdened consumers react any differently to their recession reversals than did their grandparents?

As for business, serious investment is out of the question in an atmosphere wherein many signs point to dramatic increases in taxation in the foreseeable future, as government leaders worry more about deficits than about economic growth. Moreover, the QE program promises high inflation to many in spite of assurances from the Fed that this is not a danger. These and other uncertainties confound the ability to anticipate the profitability of investments even when opportunities for them appear. They underscore the importance of a renewed technology-based investment, for it will take an inducement of that magnitude to overcome heightened uncertainties presently plaguing private investment decisions.

Is investment any more than the means by which the benefits of a technology revolution are brought to bear on the general economy? The answer is yes, but the "yes" is qualified. The problem is that what is commonly termed "investment" is applied to a huge range of expenditures whose growth potential varies widely. Short-term government countercyclical measures are typically applied without regard for the limited growth-inducing quality of such slate of investment opportunities as does exist under the very conditions under which countercyclical measures are deemed necessary.

Possibilities for the Future

The amount of future economic growth that will overcome many of today's problems will not occur with simple continuation of the anemic recovery to the late great recession. More technology revolutions are required. These will involve confluences of technologies whose implementation demands massive new production facilities and infrastructure. It is possible that some of the new technologies visible today will contribute to a future technology nexus. Here are some possibilities. There are many others.

Recombinant DNA

The modeling of the DNA molecule in the 1950s and several subsequent developments offer a highly intriguing slate of conjectural possibilities for the future. Subsequent developments include recombinant DNA, and the Human Genome project. One of the effects of these developments was the idea that the ability to create miracle cures for some of the most challenging of human diseases was a near-term prospect. The genome was fully mapped in 2000. Ten years later, the seeming promise of this technology has not been realized. While there were a small number of cases in which gene abnormalities could be linked to specific diseases, these have come to be recognized as low-hanging fruit. Simply put, the human genome and its relation to diseases has been found to be much more complex than was supposed at the time the genome was being sequenced.

Is the lack of miracle cures legitimate grounds for disappointment? If the last ten years are viewed as the opening of a stage-one innovation, then the heady optimism of the 1990s can be dismissed as over-optimism in an era in which even experts in many fields of endeavor have come to expect quick results. How long did it take for the IT revolution to develop? Individual antecedents of IT go back, in some cases, for centuries. The basic concept of the modern computer can be found before 1850 in the writings and designs of Charles Babbage. What is presently known about the human genome can justly be regarded as the result of towering but accumulated intellectual and scientific achievement. Technology will serve the future, but as yet, no one has discovered how to hurry its basic genesis.

The pattern is a familiar one. A number of small firms were formed to exploit the techniques of bioengineering in the 1970s. There were early successes with the development of seeds with desirable characteristics, but less success with the development of engineered drugs for human health care—the complexities of this kind of development rank with other early discoveries. Also, bioengineered products have encountered popular resistance, especially in agricultural applications. Bioengineered seeds are probably the best presently available solution to expanding world food supply for an expanding population, and for this reason, the resistance to the use of these seeds will probably wane. Such resistance historically has met other new technology. Railroads were once bad because locomotives frightened horses, and human beings supposedly could not survive at the lightning speeds promised—thirty miles per hour.

Shale Gas

Musings about the technology future would do well to notice a technology revolution that has quietly emerged in the last fifteen years and whose implications are only now beginning to be perceived: the shale gas revolution. This refers to the development of techniques for producing natural gas from shale formations which have long been known to contain large quantities of gas that were widely deemed uneconomical to produce as recently as 1990. This change has involved a number of elements familiar to students of past technology revolutions. First, the technology was and is being developed by a number of smaller companies. Second, the result promises to be an abundant source of natural gas that has the security and low-cost deliverability to markets that characterize domestic U.S. sources of energy. Third, the shale gas industry is beginning a period of consolidation in which international major oil companies enter the market through acquisition of established players in the shale gas industry. One thing that makes this development intensely interesting is its size: the reserves of gas appear to be sufficient for one hundred or more years of supply at recent rates of U.S. consumption.

How much of a general technology boom can emanate from self-sufficiency in fuel gas? It could be considerable. In the first place, many of the shale formations that have begun to be developed for gas potential also have liquid deposits. A recently growing disparity between the prices of oil and gas, with oil price rising relatively, has driven interest in the oil part of the shale deposits, and techniques learned while exploiting these deposits for gas are proving applicable to oil production. As with gas, the oil potential is large. These developments are well enough progressed that one can indulge in thoughts of some of the wider implications. First, what would happen if the United States were to become a net exporter of energy products? Think of what would happen to the balance of trade if the United States were to become a net exporter of energy. Moreover, think of wider geopolitical implications.

Shale gas has, in classical manner, attracted opposition. One of the central bases for such opposition is that burning gas is not a perfect elimination of greenhouse gas emissions. Inasmuch as gas combustion produces far less CO_2 than burning coal which it displaces, it promises a massive reduction in greenhouse gas production. Shale gas development has one large political advantage against the resistance it has so far encountered, mainly in the form of temporary moratoria enacted by state legislatures: the royalty system. Surface ownership in the shale gas fields includes thousands of landowners who receive royalty payments. This tends to persuade landowners to sympathy with the development, and the political process has to recognize this.

Nanotechnology

The science of nanotechnology has emerged since 1980. In the terms of this book, it is in stage-one innovation. For example, one of its more alluring potentials is in the area of low- or no-resistance power transmission, for some nanotubes have near-superconductive properties. However, laboratory research has so far not discovered how to produce these tubes separately from nanotubes that lack this property. It is to be supposed that doing this in the laboratory has to precede doing it industrially. However, the possibilities toward which superconducting and other properties point could easily underpin a major future investment boom.

On Growth Itself

Inasmuch as the contribution of a technology revolution can be a growth-inducing investment boom, it is not out of place to ask why growth itself should be such an important goal. After all, John Stuart Mill once envisioned an economy that did not grow, and as he described it, it did not sound so bad. Moreover, the idea that an economy can exist in a prosperous steady state is the implication of the diagram depicting the circular flows of incomes and expenditures that graces the opening pages of most of the popular principles of economics texts used in university instruction. Moreover, it is well known that growth has its downsides, one of which is that it magnifies income inequalities and their attendant social problems.

One of the problems surrounding a steady-state economy is that an economy of any complexity in all probability contains dynamic forces which make the attainment of conditions needed for steady-state prosperity all but impossible to attain. Perhaps the most conspicuous of these is demographics. At present, the United States faces the prospect of a retired population that is expanding relative to its working population. In this regard, the United States is far from the worst position among industrialized nations. For various reasons, the looming imbalance in Japan is worse. The same is true in China due to that country's history of draconian family-planning restrictions. Western Europe faces a similar problem. If the United States faces less of a problem than these other countries, it is because of its comparatively welcoming immigration policies, for new citizens contribute materially to a fertility rate that is very close to maintaining the balance between workers and retirees in the long term.

The problem posed by relatively growing populations of the aging is that of conflict between young and old for resources. This is because resources available in any time interval consist only of what can be produced in the same time interval; there is no intergenerational saving of resources. All the political process can do in this regard is to redistribute claims to resources as between population groups. Thus the importance of growth is simply stated. It makes the total resource base larger such that how current resources are distributed between working and retired creates less of a potential for intergenerational conflict. Foreshadowings of future conflicts of this nature can be seen now in the form of proposals to raise retirement ages and to diminish retirement benefits. Still, the best way to accommodate the needs of different generations and minimize the potential for conflict among them is to make the total economy larger.

Notes

1. For example, CNOOC's bid to buy Union Oil and Dubai World's attempt to buy U.S. port facilities were blocked by essentially political means.

2. Elementary economic theory predicts that a massive Fed purchase of U.S. securities from its member banks will lower interest rates on bonds of intermediate-to-long maturity. This logic is at the essence of the "sophisticated" models on which the Fed rests its expectations of lower rates in these classes of bonds. However, the models make little or no allowance for inflationary expectations which money printing on this scale engenders. The model treats the economy as a machine that will respond predictably to its operator's control. The initial reaction of QE, interestingly, has been an *increase* in intermediate and long interest rates, the contrary of Fed plans.

Bibliography

Books

Arthur, W. Brian. *The Nature of Technology: What it is and How it Evolves.* New York: Free Press, 2009.

Barfield, Claude. *High-tech Protectionism: The Irrationality of Antidumping Laws.* Washington, DC: AEI Press, 2003.

Burk, James. *The Pinball Effect.* New York: Little Brown, 1996.

Chamberlin, Edward H. *The Theory of Monopolistic Competition.* Cambridge: Harvard University Press, 1965.

Cummins, Lyle. *Internal Fire: The Internal Combustion Engine, 1673-1900.* 3rd ed. Wilsonville, OR: Carnot Press, 2000.

Diamond, Jared. *Guns, Germs, and Steel.* New York: W.W. Norton, 1999.

Edmonson, Nathan. *Technological Foundations of Cyclic Economic Growth: The Case of the United States Economy.* New Brunswick, NJ: Transactions Publishers, 2009.

Fishlow, A. "Productivity and Technical Change in the Railroad Sector, 1840-1910." In *Output, Employment, and Productivity in the U.S. after 1800.* Studies in Income and Wealth No. 30. New York: National Bureau of Economic Research, 1966.

Fishlow, A., and K. Parker, eds. *Growing Apart.* New York: Council on Foreign Relations Press, 1999.

Friedman, Milton, and Anna J. Schwartz. *A Monetary History of the United States, 1857-1960.* Princeton, NJ: Princeton University Press, 1963.

_____. *The Optimum Quantity of Money and Other Essays.* Chicago, IL: Aldine Publishing, 1969.

Haberler, Gottfried. *Prosperity and Depression.* 3rd ed. Lake Success, NY: United Nations, 1946.

Hayek, F. A. *The Road to Serfdom.* Chicago, IL: The University of Chicago Press, 1944.

Hicks, J. R. *A Contribution to the Theory of the Trade Cycle.* Oxford: Oxford University Press, 1950.

_____. *Value and Capital.* 2nd ed. Oxford: Oxford University Press, 1965.

Kuhn, Thomas S. *The Structure of Scientific Revolutions.* 3rd ed. Chicago, IL: University of Chicago Press, 1996.

Mandelbrot, Benoit. *Fractals and Scaling in Finance.* New York: Springer, 1997.

_____. *The Fractal Geometry of Nature.* New York: W.H. Freeman, 1983.

Mandelbrot, Benoit and Richard L. Hudson. *The (Mis)behavior of Markets.* New York: Basic Books, 2004.

Minsky, Hyman P. *Stabilizing an Unstable Economy.* New Haven, CT: Yale University Press, 1986.

Pasour, E. C., Jr., and Randal R. Rucker. *Plowshares and Pork Barrels: The Political Economy of Agriculture.* Oakland, CA: The Independent Institute, 2005.

Perez, Carlota. *Technological Revolutions and Financial Capital.* Aldershot: Edward Elgar, 2002.

Pietgen, Heinz-Otto, Hartmut Jürgens, and Ditmar Saupe. *Chaos and Fractals.* New York: Springer-Verlag, 1992.

Posner, Kenneth A. *Stalking the Black Swan.* New York: Columbia University Press, 2010.

Reinhart, Hugo, and Erik S. Reinhart. "Creative Destruction in Economics: Nietzsche, Sombart, Schumpeter, Joseph A." In *The European Heritage in Economics and the Social Sciences,* edited by Backhaus, Jürgen, and Drechsler (Boston, MA: Kluwer).

Rosenberg, Nathan. *Inside the Black Box: Technology and Economics.* Cambridge: Cambridge University Press, 1982.

Schlaes, Amity. *The Forgotten Man.* New York: Harper Collins, 2007.

Schumpeter, Joseph A. "The Analysis of Economic Change," *The Review of Economic Statistics* (May 1935) (as reprinted in American Economic Association, *Readings in Business Cycle Theory.* Homewood: Richard D. Irwin, 1951), 1–19.

_____. *Capitalism, Socialism, and Democracy.* 3rd ed. New York: Harper Perennial, 1950.

_____. *The European Heritage in Economics and the Social Sciences.* Edited by Backhaus, Jürgen, and Drechsler (Boston, MA: Kluwer).

_____. *The Theory of Economic Development.* Reprint of the 1934 translation from the German by Harvard University Press. New Brunswick, NJ: Transaction Publishers, 2005.

Smith, Vernon L. *Investment and Production.* Cambridge: Harvard University Press, 1961.

Taleb, Nicholas Nassim. *The Black Swan.* New York: Random House, 2007.

Triplett, Jack E., and Barry P. Bosworth. *Productivity in the U.S. Service Sector.* Washington, DC: Brookings Institution, 2004.

Usher, Abbott Payson. *A History of Mechanical Inventions.* Boston, MA: Beacon Press, 1929.

Articles in Periodicals

"Buys and Modernizes Steam Engines," *Modern Railroads* 4, no. 11 (November 1949): 81–86.

Barro, Robert J. "Demand Side Voodoo Economics." *The Economists Voice* 6, no. 2 (2009): Article 5.

Feldstein, Martin S., and David K. Foot. "The Other Half of Gross Investment: Replacement and Modernization Expenditures." *Review of Economics and Statistics* 53, no. 1 (February 1971): 49–58.

Feldstein, Martin S., and Michael Rothschild. "Towards an Economic Theory of Replacement Investment." *Econometrica* 42 (May 1974): 393–416.

Solow, Robert M. "Technical Change and the Aggregate Production Function." *Review of Economics and Statistics* 39 (1957).

Van Zandweghe, Willem. "Why Have the Dynamics of Labor Productivity Changed?" *Economic Review* 95, no. 3 (3rd quarter, 2010).

Wald, Mathew L. "Giant Holes in the Ground." *Technology Review* 113, no. 6 (December 2010): 60–65.

Zweibel, Ken, James Mason, and Vasilis Fthenakis. "A Solar Grand Plan." *Scientific American* 294 (January 2008): 64–73.

Other Sources

Gillis, Malcolm, "New Perspectives on 21st Century Technology: The Nano-Bio-Info Convergence," (Rice University, Department of Economics). Prepared for presentation at Federal Reserve Bank of Dallas, Houston Branch, Apr. 20, 2010.

Isaac, William N. *Testimony Before the House Subcommittee on Capital Markets, Insurance, and Government-Sponsored Enterprises*, March 12, 2009.

Index